Field Trials in Oil Palm Breeding

A Manual

Techniques in Plantation Science Series

Series Editors:

Brian P. Forster, Lead Scientist, Verdant Bioscience, Indonesia
Peter D.S. Caligari, Science Strategy Director, Verdant Bioscience, Indonesia

About the series:

A series of manuals covering techniques in plantation science that form the essential underlying needs to carry out plantation work of this kind.

The series reflects the expertise in Verdant Bioscience that underlies the plantation science activities carried out at the Verdant Plantation Science Centre (VPSC) at Timbang Deli, Deli Serdang, North Sumatra, Indonesia.

Titles available:

Field Trials in Oil Palm Breeding

A Manual

Baihaqi Sitepu
Verdant Bioscience, Indonesia

Nur Dian Laksono
Verdant Bioscience, Indonesia

Umi Setiawati
Verdant Bioscience, Indonesia

Fazrin Nur
Verdant Bioscience, Indonesia

Miranti Rahmaningsih
Verdant Bioscience, Indonesia

Yassier Anwar
Verdant Bioscience, Indonesia

Pujo Widodo
Verdant Bioscience, Indonesia

Brian P. Forster
Verdant Bioscience, Indonesia

Abdul R. Purba
Indonesian Oil Palm Research Institute, Indonesia

CABI is a trading name of CAB International

CABI	CABI
Nosworthy Way	745 Atlantic Avenue
Wallingford	8th Floor
Oxfordshire OX10 8DE	Boston, MA 02111
UK	USA

Tel: +44 (0)1491 832111
Fax: +44 (0)1491 833508
E-mail: info@cabi.org
Website: www.cabi.org

Tel: +1 (617)682-9015
E-mail: cabi-nao@cabi.org

A catalogue record for this book is available from the British Library, London, UK.

Library of Congress Cataloging-in-Publication Data

Names: Sitepu, Baihaqi, author.
Title: Field trials in oil palm breeding : a manual / Baihaqi Sitepu, Verdant Bioscience, Indonesia, Nur D. Laksono, Verdant Bioscience, Indonesia, Umi Setiawati, Verdant Bioscience, Indonesia, Fazrin Nur, Verdant Bioscience, Indonesia, Miranti Rahmaningsih, Verdant Bioscience, Indonesia, Yassier Anwar, Verdant Bioscience, Indonesia, Pujo Widodo, Verdant Bioscience, Indonesia, Brian P. Forster, Verdant Bioscience, Indonesia, Abdul R. Purba, Indonesian Oil Palm Research Institute, Indonesia.
Description: Wallingford, Oxfordshire, UK ; Boston, MA : CABI, [2020] | Series: Techniques in plantation science | Includes bibliographical references and index. | Summary: "This is a hands-on, practical guide to describe field trials in oil palm. The location for field trials is key, as is land preparation"-- Provided by publisher.
Identifiers: LCCN 2019030925 (print) | LCCN 2019030926 (ebook) | ISBN 9781789241396 (paperback) | ISBN 9781789241402 (ebook) | ISBN 9781789241419 (epub)
Subjects: LCSH: Oil palm--Breeding.
Classification: LCC SB299.P3 S57 2020 (print) | LCC SB299.P3 (ebook) | DDC 641.3/3851--dc23
LC record available at https://lccn.loc.gov/2019030925
LC ebook record available at https://lccn.loc.gov/2019030926

ISBN-13: 978 1 78924 139 6 (paperback)
        978 1 78924 140 2 (ePDF)
        978 1 78924 141 9 (ePub)

Commissioning Editor: Rebecca Stubbs
Editorial Assistant: Emma McCann
Production Editor: James Bishop

Typeset by SPi, Pondicherry, India
Printed and bound in the UK by Severn, Gloucester

Series Foreword – Techniques in Plantation Science

Verdant Bioscience, Singapore (VBS) is a company established in October 2013 with a vision to develop high-yielding, high-quality planting material in oil palm and rubber through the application of sound practices based on scientific innovation in plant breeding. The approach is to fuse traditional breeding strategies with the latest methods in biotechnology. These techniques are integrated with expertise and the application of sustainable aspects of agronomy and crop protection, alongside information and imaging technology which not only find relevance in direct aspects of plantation practice but also in selection within the breeding programme. When high-yielding planting material is allied with efficient plantation practices, it leads to what may be termed 'intensive sustainable' production. At the same time, the quality of new products is refined to give more specialised uses alongside more commodity-based oil production, thus meeting the market demands of the modern world community, but with a minimal harmful footprint. An essential ingredient in all this is having sound and practical protocols and techniques to allow the realisation of the strategies that are envisaged.

To achieve its aims, VBS acquired an Indonesian company called PT Timbang Deli Indonesia, with an estate of over 970 ha of land at Timbang Deli, Deli Serdang, North Sumatra, Indonesia, and the group works under the name of 'Verdant'. A central part of this estate, which will be used for important plant nurseries and field trials, is the development of the Verdant Plantation Science Centre (VPSC), to which the operational staff moved in October 2016. A seed production and marketing facility is now established at VPSC for commercial seed sales and the processing of seed from breeding programmes. The centre comprises specialised laboratories in cell biology, genomics, tissue culture, pollen, soil DNA, plant and soil nutrition, bunch and oil, agronomy and crop protection. Field facilities include extensive nurseries, seed gardens and trials (trial sites are also located at various places across Indonesia). It is the aim of the company to use its existing and rapidly developing intellectual property (IP) to develop superior cultivars that not only have outstanding yield but are also resistant to both biotic and abiotic stresses, while at the same time meeting new market demands. Verdant not only develops and supplies superior planting materials but also supports its customers and growers with a package of services and advice in fertiliser recommendations and crop protection. This is all part of a central mission to promote green, eco-friendly agriculture.

Brian P. Forster and Peter D.S. Caligari
Lead Scientist and Science Strategy Director
Verdant Bioscience

Contents

Acknowledgements

The authors are grateful to all the breeding and biotechnology teams of Verdant for sharing their knowledge and providing helpful advice in preparing this manual.

Preface

As noted in the foreword to this series, a central objective in Verdant's mission is to develop better, more productive and more sustainable cultivars of oil palm, rubber and other plantation crops, particularly through plant breeding. Field testing is an essential component in selecting and developing new varieties with superior performance for yield, quality, pest and disease resistance and in meeting new market demands – for example, specialised oil quality. This manual covers the basic and changing practices such as land preparation, planting and data recording of trials. Field trialling of oil palm includes breeding trials (selection within promising new genotypic combinations or germplasm), particularly by progeny testing to establish new commercial lines and identifying *Ganoderma* resistant/tolerant material (*Ganoderma* is the most important disease of oil palm in Southeast Asia, while *Fusarium* is in Africa). Typically, palms are grown in the field and their performance is monitored, effectively meaning that selection is largely based on phenotype. Oil palm breeding is now moving into the genomics era and selection based on genotype will play an increasingly important role – for example, DNA markers are now available for important traits such as shell thickness and fruit colour. This means that these traits can be selected prior to field trialling, thus saving space, time and costs. However, the 'acid test' is, and always will be, performance in the field. The efficiency and effectiveness of the selection will be seen in the improvements in the resulting cultivars which, as they are adopted by commercial plantations, will help in sustainability goals by giving more oil product per planted ha. This, combined with better disease resistance, will mean more yield from existing areas and less pressure for new land. This manual forms part of a series in 'Techniques in Plantation Science', and fits 'in between' *Nursery Practices in Oil Palm* and *Seed Production in Oil Palm*. It follows on from *Crossing in Oil Palm*. Our target audiences are plant breeders, planters, students and researchers in oil palm agriculture, along with plant breeders and end-users interested in the practicalities of producing high quality oil palm planting materials for breeding and commercial production.

Brian P. Forster and Peter D.S. Caligari

Introduction

1

Abstract

Field testing – trialling – of oil palm is subject to many factors, chief among them being the basic biological features of the crop (it is a long-lived perennial) and the traits to be assessed (mainly yield, thus mature palms need to be produced). Trialling is the next step after breeders have produced progeny from deliberate crosses, and this in turn is dependent upon the genetic variation available to the breeder (see Setiawati *et al.*, 2018, this series). Trialling allows the selection of progenies and palms based on field performance, and these can then be promoted to variety status and commercial production (see Kelanaputra *et al.*, 2018, this series). Oil palm, *Elaeis guineensis*, is the world's most important oil crop. Although it has been used to some extent as a food since ancient times in its centre of origin (West Africa), it is a relatively modern crop, with Southeast Asia now being the main production region. A brief history of the development of the oil palm crop (significant dates and eras) is given. Improvement of the crop through breeding and trialling is relatively recent, with beginnings in the early 20th century. Since breeding is dependent upon access to traits of interest, germplasm collections from the centre of origin have been made. In addition, variation is being produced by mutation induction and selection (see Nur *et al.*, 2018, this series). Thus, oil palm breeders have access to a rich germplasm base and can create progenies which combine novel traits (e.g. disease resistance) with traditional traits (e.g. thin-shelled fruit type), which then need to be tested in field trials.

1.1 History of Oil Palm Cultivation and Crop Facts

Oil palm (*Elaeis guineensis* Jacq) is a species of the humid tropics; the centre of diversity of the species is West Africa, but semi-wild and cultivated forms are found in other parts of Africa, Southeast Asia and South America.

Oil palm is the highest yielding of all oil-bearing crop plants – for example, oil palm produces 11 times higher yields per land area than soybean, the second highest oil crop plant. Oil is extracted from the oil palm fruit, from both the fleshy mesocarp and the kernel.

Oil palm's Latin genus name is derived from the Greek word *elaion* meaning 'oil' and the species name indicates its West African origins in and around Guinea. The crop first came to the attention of the outside world due to travellers to Africa in the 15th century (see Corley and Tinker, 2015 for more details on the history of oil palm). The first plantings in Indonesia, which led to its rise in importance to become the world's top oil crop, did not occur until the late 19th century. Large-scale plantations were established in the early 20th century in both Africa and Southeast Asia. Initial plantations were composed of dura palms which are characterised as having thick-shelled fruits (Fig. 1.1a). In the 1920s the first crosses were made in deliberate attempts to improve the crop through plant breeding, and in the 1950s–1960s the more productive tenera types took over as the favoured commercial material both in Africa and Southeast Asia. Tenera genotypes are thin-shelled (Fig. 1.1b) and have thick oil-bearing fruit flesh that yields 30% more oil than dura fruit forms. Tenera (thin-shelled) types are produced from deliberate crossing of dura (thick-shelled) and pisifera (no shell) parents (Fig. 1.1). Thus, artificial crossing became an essential and major component in commercial oil palm seed production as well as in breeding.

Indonesia is currently the main producer of palm oil. Cultivation in Indonesia started when four oil palm seedlings from Mauritius and the Netherlands were planted in the Bogor Botanical Garden in Java in 1848. Ten years after these seedlings were planted, seeds were distributed to other areas of Java, and to the neighbouring island of Sumatra, but no industrial plantation was established until 1911, when Belgian and German

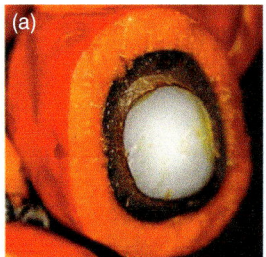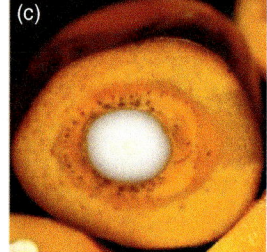

Fig. 1.1. Anatomy of fruit forms of oil palm. a) Dura fruits have a thick-shelled nut ('dura' is derived from the Latin meaning 'hard/thick/tough'); b) Tenera fruits have a thin-shelled nut surrounded by a fibre ring ('tenera' is derived from the Latin meaning 'soft/tender/weak'); c) Pisifera fruits have a naked, no-shell kernel surrounded by traces of a fibre ring ('pisifera' is a description of the kernel, derived from the Latin meaning pea-like).

companies planted oil palm at Sungai Liput, Pulu Radja and Tanah Itam Ulu, in Sumatra, Indonesia. From this basis, the oil palm industry grew in Indonesia to a planted area of 2,760 ha in 1915, which had increased to 12,307,677 ha by 2018.

Oil palm was introduced to Malaysia in 1870, through the Botanical Garden of Singapore, as an ornamental plant (these were probably derived from palms from the Bogor Botanical Garden). The Malaysian oil palm industry commenced around 1917, at a similar time to that of Indonesia. A distinct new era began in Malaysia in 1960 when the rubber price fell and plantations moved into oil palm as a more profitable alternative. The oil palm plantation area grew from 31,000 ha in 1960 to almost 600,000 ha in 1980 (Pamin, 1998).

A comprehensive review of the oil palm crop is given by Corley and Tinker (2015), and only a brief description is given here. Oil palm is grown in the humid tropics between 20° latitudes north and south of the equator. The crop covers over 8.5 million ha worldwide. Oil palm is highly profitable and grown both on large-scale plantations and by smallholders (Sayer *et al.*, 2012). Ripe fruit bunches are harvested continually and sent to local mills for oil extraction. Oil palm fruits provide both crude palm oil (CPO) and palm kernel oil (PKO), extracted from the fruit flesh (mesocarp) and kernel (endosperm) respectively. CPO is made up of mainly palmitic (43%), oleic (39%), stearic (5%) and other fatty acids (Siew, 2002), and is a major source of pro-vitamin A and vitamin E (Barcelos *et al.*, 2015). PKO is a high-quality oil containing lauric (up to 50%), myristic (15%) and other essential fatty acids (Sambanthamurthi *et al.*, 2000). Since oil palm is harvested continually, CPO represents a relatively stable commodity compared to annual oil crops. The main CPO producing countries are Indonesia (53% of global production) and Malaysia (38%); the largest consumers are India (28% of the market), China (22%) and Europe (22%).

The future demand and supply balance in the oil palm industry is not easy to foresee. Demand is expected to continue to increase because of the expansion of the human population and increasing living standards (see Jalani, 1988). This trend may impact on the main markets for oil palm products, with China, India and Pakistan becoming major importers and consumers (Mielke, 2001). Estimates of the increasing demands for foodstuff in the next 30 years are usually expressed in terms of cereal demand, and this is expected to increase by roughly 50% by 2030. The demand for fats/oils will be expected to increase proportionally more rapidly, and Mielke (2001) predicted that the demand for oil palm would double by 2020 compared to 2001. Palm oil is sold mainly as a commodity, CPO, but it is likely that breeding will introduce new, novel oil quality traits for specialised end-users and adapted to changing consumer demands – for example, biofuels and bioplastics. Thus, quality traits and new products are expected to impact on future markets.

1.2   History of Oil Palm Breeding

Current oil palm breeding primarily aims to maximise oil and kernel yield, and thus contributes to plantation profitability. The breeder therefore aims to select for high yields of fruit and high kernel oil content and make crosses between the best individuals. Subsidiary objectives may include reduced height (for ease in harvesting) and resistance/tolerance to diseases and stresses. To date there has been very little selection for oil quality, though this is expected to change in the future (Chapter 7). In order to make progress, the breeder must start with a population of palms in which there is genetic variation for yield (and other target traits). Starting with a variable population, the breeder must then decide which characteristics to select for, and this is dependent upon trialling (Selection criteria and methods of measurement will be discussed in Chapters 6 and 7.) It is relatively easy to make improvements from one generation to the next (selection progress) for characteristics that are highly variable, and for which much of the variation is controlled by genetic (heritable) differences, and less by environmental factors. However, for some traits (e.g. resistance to *Ganoderma*, a cause of major oil palm fungal disease), there is no or little variation in the primary gene pool and the breeder may be forced to introduce desirable traits from wild or semi-wild gene pools. However, this would involve a long, protracted breeding scheme. Accelerated breeding methods may be adopted in such cases. For example, mutation breeding (Nur *et al.*, 2018), transformation (Masli *et al.*, 2009; Masani *et al.*, 2018) or gene editing (Murphy, 2018).

The aim of present-day oil palm breeding programmes is the development of thin-shelled tenera (commercial) palms. Shell thickness is controlled by a single gene, *Sh* (Beirnaert and Vanderweyen, 1941; Singh *et al.*, 2013; Chapter 9). Earlier breeding programmes were largely based on phenotypic mass selection combined with family selection in bi-parental crosses. Today a primary objective is the production of duras (mother palms) which when crossed in combination with pisiferas (male parents) produce high yielding teneras (see Setiawati *et al.*, 2018, this series). Female sterility of pisiferas determines the direction of these crosses, i.e. pisiferas are used as male (pollen) parents, and because the heterozygote is the shell type for exploitation, it necessitates progeny testing programmes for which various forms of recurrent selection may be adopted (Hardon *et al.*, 1976a).

Although oil palm has been used by man in West Africa since ancient times, its commercial exploitation is relatively recent as a world crop. With the realisation of its economic potential, concerted efforts towards oil palm improvement began at the turn of the 20th century in West and Central Africa. An account of the history of the development of the various breeding programmes in Africa has been described by Hartley (1988). These early efforts include the work of the Institut national pour l'étude agronomique du Congo belge (INÉAC), and two major West African institutes: the West African

Institute for Oil Palm Research (WAIFOR based in Nigeria, Ghana and Sierra Leone) and the Institute de Recerche pour Les Huiles et Oleagineux (IRHO) in Cote d'Ivoire and Benin.

The work in Nigeria was started by E.H.G Smith of the Nigerian Department of Agriculture by exploiting the genetic variation in oil palm populations from Calabar, Aba, Nkwele (Umuahia) and later Ufuma in natural groves located in eastern Nigeria regions. Smith planted 800 oil palms at Calabar in south-eastern Nigeria. Yield and bunch data of these plants (arising from seed of open-pollinated bunches collected from local wild palm groves) were collated from 1922–1928. From this, Smith selected nine duras (fruit with thick-shelled kernels) and ten teneras (thin-shelled). Twelve of the selected palms were self-pollinated to form the Calabar hybrid generation and the progenies were planted in four breeding stations of the Department of Agriculture (Ogba, Umudike, Ibadan and Nkwelle) between 1930 and 1935. These stations supplied seed for extension work and for experimentation and field trialling (Barbosa and Chinchilla, 2003).

The oil palm industry in Indonesia has been promoted through research and development. A major step forward was the establishment of the Algemeene Proefstation der AVROS (APA), a research station of AVROS (Algemeene Vereniging van Rubber-planters ter Oostkust van Sumatra) in Medan, North Sumatra, in September 1916. While the primary intention of APA was to conduct investigations into rubber, its research also evaluated oil palm. In 1922, Rutgers compiled investigations on oil palm that had been conducted by AVROS and concluded that oil palm planting in Sumatra was very successful. The variety planted (tenera), performed better than the dura crop of West Africa, giving earlier and higher yields, but Rutgers noted that further improvement was possible through selection (see Pamin, 1998). It is believed that the material tested originated from the four palms at the Bogor Botanical Garden, Java, Indonesia (received from Africa via Mauritius and the Netherlands). Seeds from these four palms were widely distributed throughout Indonesia and planted as ornamentals (often used to line roads in rubber plantations), but later used to supply oil palm estates (from 1911 onwards). Selection programmes started at various centres in Indonesia from the 1920s and gave rise to breeding a population of duras generally referred as 'Deli dura' (Hardon and Turner, 1967) (the 'Deli' name refers to an old region in North Sumatra, Indonesia). A detailed history of the Deli dura breeding population is given in Corley and Tinker (2015).

The superior oil content of tenera (T) palms led to the distribution of TxT seed from commercial plantings in Congo in the 1930s, but by 1938 it was known that 25% of the progeny were female sterile (Beirnaert, 1940). Beirnaert explained clearly the inheritance of the shell-thickness character (a single gene trait), and brought forward evidence against the Congo theory, current in French West Africa at that time, that the tenera fruit type was a degenerate shell form. Beirnaert explained that to prevent sterility,

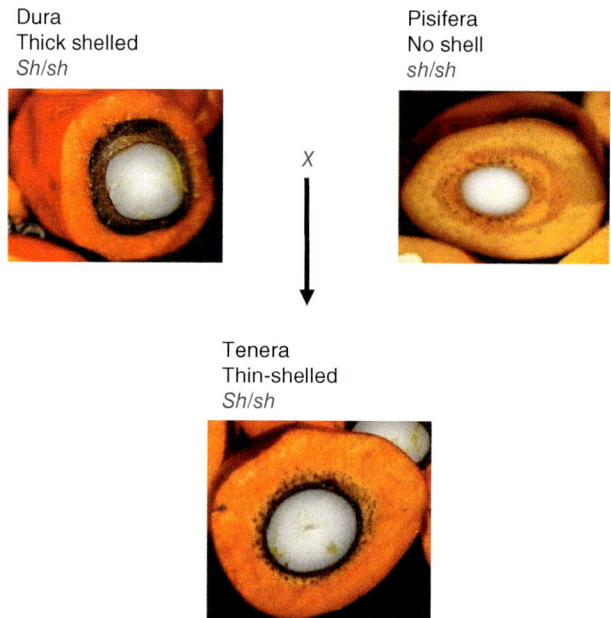

Fig. 1.2. Crossing scheme of dura (female) with pisifera (male) to produce the commercial thin-shelled tenera fruit type.

dura x tenera (DxT or TxD) should replace TxT production. A subsequent Calabar tenera self, and crosses in Nigeria, quickly showed this to be the case (Hartley, 1957). Dura x pisifera (DxP) crosses including Deli duras as parents were seen to provide 100% tenera progeny and this type of cross became standard practice. The genetic structure of these crosses is given in Fig. 1.2. The shell thickness gene, *Sh*, is semi-dominant with dura (*Sh/Sh*) being thick-shelled, pisifera (*sh/sh*) having no shell and tenera (*Sh/sh*) being intermediate/thin-shelled.

1.3   Germplasm

Parental selection for breeding and commercial production has been mostly empirical due to a deficiency of knowledge of oil palm genetics. The narrow origin of parental populations has set limits on improvement schemes that can be achieved, and it is generally accepted that further improvement will require a widening of the genetic base of both dura and pisifera germplasm, either by intercrossing or by introgressing with new introductions from un-related and wild material (Hardon *et al.*, 1973), or by mutation breeding (Nur *et al.*, 2018). Transformation and gene editing are also possibilities; however, genetically modified (GMO) oil palm is currently unacceptable on a consumer basis.

Commercial seed production is based on DxP crosses, which produces the desired thin-shelled tenera fruit. The predominant parental lines in Indonesia, Malaysia and Papua New Guinea are Deli duras (females) and AVROS pisiferas (males). Crop improvement through breeding has been limited by the restricted genetic variation contained in these elite parental gene pools. In order to make progress in breeding, it became necessary to provide breeders with more genetic variation and thus germplasm collections from wild, landrace and cultivated materials in West Africa (the centre of origin and diversity) have been carried out. WAIFOR, in Nigeria, was one of the first to do this. In recent years, major oil palm breeding companies have joined collecting expeditions in West Africa (e.g. Oykere-Boateng *et al.*, 2008; Sapey *et al.*, 2012), and more recently there have been germplasm collections in Nigeria (following the earlier collection carried out by Malaysia) and consortia have been set up to bring into Indonesia interspecific hybrids between *E. oleifera* and *E. guineensis* from South America.

The first germplasm collections in Nigeria were small and limited (60 ha). They included groves at Aba (11 ha) and Ufuma (49 ha), as well as Calabar plot materials. In the early 1960s a collection in grove areas in eastern Nigeria was undertaken, and 72 open-pollinated progenies were established at Nigeria Institute for Oil Palm Research (NIFOR). International collaborative research was established and continues to date. Collections have been carried out in Cote d'Ivoire at Yacoboué and Sasandra, and in Cameroon in the Widikum region by the IRHO (Corley and Tinker, 2015), and a prospection of the oil palm genetic material within and beyond the major oil palm belt of Nigeria has been set up between Indonesia and Nigeria for genetic conservation and exploitation.

Beginning in the 1970s, the acquisition of new germplasm was conducted by the Marihat Research Station, North Sumatra, Indonesia, in collaboration with IRHO. Introductions into Indonesia came from Ghana, Ivory Coast, Cote d'Ivoire, DR Congo, Angola, Nigeria and Papua New Guinea, and the pollen of *Elaeis oleifera* (a related S. American species) was introduced from Colombia (Pamin, 1998).

Other collections were carried out by the Malaysian Palm Oil Board (MPOB) in oil palm's centre of origin in order to broaden the genetic base of oil palm planting material. Countries in Africa in which oil palm was explored included Nigeria, Cameroon, DR Congo, Tanzania, Madagascar, Angola, Senegal, Gambia, Sierra Leone, Guinea and Ghana. A total of 723 accessions were collected to widen the oil palm germplasm for breeding and research purposes in Malaysia (Hayati *et al.*, 2004).

Oil palm germplasm is conserved mainly as living palm trees, and to a lesser extent as pollen. The former requires large land areas but may be maintained for decades (some collections have trees more than 100 years old). Pollen, however may be stored for up to 20 years vacuum-packed in −20°C freezers. Another approach suggested for oil palm is to cryopreserve germplasm at −198°C in liquid nitrogen (Grout *et al.*, 1983).

1.4   Traits

The basic biology of oil palm has direct effects on the traits of interest and how breeding, harvesting and trialling are conducted. Oil palm is a long-lived plant with many fascinating aspects relating to vegetative and reproductive biology, ecology, morphology and agronomy. The large size and long generation time of oil palm creates many formidable challenges for researchers and breeders, especially in comparison with the much smaller and shorter generation time of annual oilseed crops such as soybean, rapeseed and sunflower. It is therefore essential that oil palm improvement programmes are focused on a limited number of key target traits. By far the most important trait is oil yield, followed by disease resistance (especially *Ganoderma* in Indonesia and *Fusarium* in Africa), oil quality and composition, and tolerance to a range of pests. Currently, oil palm is harvested by hand, but several traits (e.g. long stem and retention of fruits) may be combined to adapt the crop to mechanical harvesting.

As with most crops, yield is the most important trait for oil palm breeding. Yield (but really oil yield) per land area is of particular importance as the oil palm industry is under pressure to be more environmentally friendly and sustainable, by limiting expansion to conserve rainforest biodiversity, thus increasing yield per land area is a major target. Some target traits for oil palm improvement have been reviewed by Forster *et al.* (2018), they include:

- **fruit type:** fruit type is based on the thickness, presence or absence of the shell covering the kernel, the presence of a fibre ring and the thickness of the oil-bearing fruit flesh (mesocarp)
- **short stature:** reduced height enables longer plantation life
- **frond length:** reduced frond length allows greater planting densities
- **precocity:** early flowering brings early economic returns
- **resistance to disease:** wilt in Africa and *Ganoderma* in Southeast Asia
- **mechanised harvesting:** ripe fruit colour, long bunch stalk, fruit retention
- **oil quality:** to the allow development of specialised oils for specific end users
- **non-food uses:** cosmetics, biofuel, industrial.

It is difficult to determine future market demands as they change continually. In addition to the traits listed above there are now new eco-friendly opportunities for oil palm such as the production of biodiesel and bioplastics.

1.5   Trialling

The breeding, selection and evaluation of improved crop varieties are basic pursuits in which trialling plays a key role. The Nobel Peace Prize laureate for 1970, Norman Borlaug, made the important point that in order to achieve crop improvement, both conventional breeding and biotechnology methodologies are needed, but that in the end everything has to be tested in the field. Field trialling is therefore vital, as are statistical principles in trial design and data analysis as these bring rigour in evaluating field performance (Smith *et al.*, 2005). The breeding process starts with identifying the trait(s) to be improved. Desirable variation is then sought in available germplasm and crosses made to initiate the introgression of the trait into elite material (practical methods in crossing in oil palm are given in Setiawati *et al.*, 2018). The progenies are then tested at various stages in development (from seed, to germination and seedling traits, through to maturity) for which trialling is essential. Trials allow the selection of superior lines, having improved traits that exhibit stability. These are then advanced for multiplication, official progeny testing and varietal release. The practicalities of field trialling are laid out in subsequent chapters.

References

Barbosa, R. and Chinchilla, C. (2003) ASD oil palm germplasm from Nigeria. *ASD Oil Palm Papers* 26: 33–44.

Barcelos, E., Rios, S.A, Cunha, R.N., Lopes, R., Motoike, S.Y., Babiychuk, E., Skirycz, A. and Kushnir, S. (2015) Oil palm natural diversity and the potential for yield improvement. *Frontiers in Plant Science* 6: 190.

Beirnaert A. (1940) Le problème de la sterilité chez le palmier à huile. *Bull. Agric, Congo Belge* 3: 95.

Beirnaert A. and Vanderweyen R. (1941) Contribution a l'étude génétique et biométrique des variétés d'Elaeis guineensis Jacquin. *Publ. Inst. Nat. Etude Agron, Congo Belge Ser. Sci.* 27:1–101.

Corley, R.H.V. and Tinker, P.B. (2015) *The Oil Palm*, 5th edn. World Agriculture Series. Wiley Blackwell, Hoboken, NJ, p. 639.

Forster, B.P., Sitepu, B., Setiawati, U., Kelanaputra, E.S., Nur, F. *et al.* (2018) Oil palm breeding. Campos, H.A. and Caligari, P.D.S. (eds). *Genetic Improvement of Tropical Species*. Springer, New York, NY, pp. 241–290.

Grout, B.W.W., Shelton, K. and Pritchard, H.W. (1983) Orthodox behaviour of oil palm seed and cryopreservation of the excised embryo for genetic conservation. *Annal of Botany* 52: 381–384.

Hardon, J.J. and Turner, P.D. (1967) Observations on natural pollination in commercial plantings of oil palm (*Elaeis guineensis*). *Malaya. Expl. Agric.* 3: 105–116.

Hardon, J.J., Mokhtar, H. and Ooi, S.C. (1973) Oil palm breeding: a review. Wastie. R.L. and Earp, D.A. (eds). Advances in oil palm cultivation. The proceedings of the International Oil Palm Conference, Kuala Lumpur, November, 1972. Incorp. Soc. Planters, Kuala Lumpur, pp. 23–87.

Hardon, J.J., Gascon, J.P., Noiret, J.M., Meunier, J., Tan, G.Y. and Tam, T.K. (1976a) Oil palm breeding – introduction in oil palm research. *Development in Crop Science* 89–108.

Hardon, J.J., Gascon, J.P., Noiret, J.M., Meunier, J., Tan, G.Y. and Tam, T.K. (1976b) Major oil palm breeding programmes in oil palm research. *Development in Crop Science* 109–124.

Hartley, C.W.S. (1957) Oil palm breeding and selection in Nigeria. *J. West Afr. Inst. Oil Palm Res.* 2: 108–115.

Hartley C.W.S. (1988) *The Oil Palm*, 3rd edn. Longman, London/New York.

Hayati, A., Wickneswari, R., Maizura, I. and Rajanaidu, N. (2004) Genetic diversity of oil palm (*Elaeis guineensis* Jacq.) germplasm collections from Africa: implication for improvement and conservation of genetic resource. *Theoretical and Applied Genetics* 108: 1274–1284.

Jalani, B.S. (1988) Research and development of oil palm towards the next millennium. A. Jatmika (ed.). International Oil Palm Conference, Bali, 1998, Paper GL/09.

Kelanaputra, E.S., Nelson, S.P.C., Setiawati, U., Sitepu, B., Nur, F. *et al.* (2018) *Seed Production in Oil Palm: A Manual. Techniques in Plantation Science.* Forster, B.P. and Caligari, P.D.S. (eds). CAB International, Wallingford, UK, p. 59.

Masani, M.Y.A., Izawati, A.M.D., Rasid, O.A. and Parveez, G.K.A. (2018). Biotechnology of oil palm: current status of oil palm genetic transformation. *Biocatalysis and Agricultural Biotechnology* 15: 335–347.

Masli, D.I.A., Kadir, A.P.G. and Yunus, A.M.M. (2009) Transformation of oil palm using *Agrobacterium tumefaciens. Oil Palm Research* 2: 643–652.

Mielke, T. (2001) Can production of oil and fats keep pace with future consumption? Paper presented at ISF World Seed Congress, Berlin.

Murphy, D.J. (2018) Advances in the genetic modification of oil palm. Rival, A. (ed.). *Achieving Sustainable Cultivation of Oil Palm, vol. 1: Introduction, Breeding and Cultivation Techniques.* Burleigh Dodds Science Publishing, Cambridge, UK. p. 306.

Nur, F., Forster, B.P., Osei, S.A., Amiteye, S., Ciomas, J., Hoeman, S. and Jankuloski, L. (2018) *Mutation Breeding in Oil Palm: A Manual. Techniques in Plantation Science.* Forster, B.P. and Caligari, P.D.S. (eds). CAB International, Wallingford, UK, p. 63.

Okyere-Boateng, G., Dwarko, D.A., Kaledzi, P.D. and Nuerty, B.N. (2008) Collection, conservation and evaluation of the disappearing oil palm (*Elaeis guineensis* J) landraces in Ghana. *International J Pure Appl Sci* 1: 18–31.

Pamin, K. (1998) A hundred and fifty years of oil palm development in Indonesia: from the Bogor Botanical Garden to the industry. *International Oil Palm Conference*, Bali.

Sambanthamurthi, R., Kalyana, S. and Tan, Y. (2000) Chemistry and biochemistry of palm oil. *Progress in Lipid Research* 39: 507–558.

Sapey, E., Adusei-Fosu, K., Agyei-Dwarko, D. and Okyere-Boateng, G. (2012) Collection of oil palm (*Elaeis guineensis* Jacq) germplasm in the northern region of Ghana. *Asian Journal of Agricultural Sciences* 4: 325–328.

Sayer, J., Ghazoul, J., Nelson, P. and Klintuni, B.A. (2012) Oil palm expansion transforms tropical landscapes and livelihoods. *Global Food Secur* 1: 114–119.

Setiawati, U., Sitepu, B., Nur, F., Forster, B.P. and Dery, S. (2018) *Crossing in Oil Palm: A Manual. Techniques in Plantation Science.* Forster, B.P. and Caligari, P.D.S. (eds). CAB International, Wallingford, UK, p. 50.

Siew, W.L. (2002) Palm Oil. Gunstone F.D. (ed.). *Vegetable Oil in Food Technology: Composition, Properties and Uses.* Wiley-Blackwell, Hoboken, NJ, pp. 25–58.

Singh, R., Low, E.T.L., Ooi, L.C.L., Ong-Abdullah, M., Ting, N.C., Nagappan, J. *et al.* (2013) The oil palm SHELL gene controls oil yield and encodes a homologue of SEEDSTICK. *Nature* 500: 340–344.

Smith, A.B, Cullis, B.R. and Thompson, R. (2005) The analysis of crop cultivar breeding and evaluation trials: an overview of current mixed model approaches. *Journal of Agricultural Science* 143: 449–462. Cambridge University Press, Cambridge, UK.

Health and Safety Considerations   **2**

Abstract

Standard health and safety protocols are needed in all breeding trial activities, and these include activities in the field, nursery and the laboratory. The official standards can vary depending on country and regulations, but good standards should be maintained whatever requirements are officially needed. Identification and elimination of hazards and risks, followed by the development of specific safety practices and procedures for preventing and responding to workplace accidents and injuries are important in establishing effective occupational health and safe working conditions. Guidelines in health and safety issues relating to field trials and related nursery and laboratory activities in oil palm breeding are given below. They are also provided by various organisations.

2.1   Health and Safety in the Field

The main activities in trialling are carried out in the field. Health and safety of workers and the environment are paramount. Work risks in the field can be high if standard health and safety procedures are not observed. Hazards vary depending on the site and the activity.

Occupational health and safety issues in the field are described by the International Labour Organization (e.g. IFC, 2016). They include the following:

- Physical hazards:
    - operational and workplace hazards
    - machinery and vehicles
    - entry into confined and restricted spaces
    - risk of fire and explosion.
- Biological hazards.
- Chemical hazards.

The IFC guidelines are available online and are revised regularly.

Personal protective equipment (PPE) is essential in protecting workers from injuries and health threats. PPE for field activities are as follows:

- Wide-rimed hats – these should be worn to provide sun protection.
- Sun cream – to avoid sunburn.
- Safety boots – these should be worn in every activity in the field due to various hazards, for example, uneven ground, spines as well as biting, stinging and nuisance insects and animals.
- Gloves – these need to be worn in some activities in the field, such as planting, bagging, crossing, spraying, marking and harvesting, to avoid bruising, abrasions, lacerations and skin contact with hazardous chemicals
- Helmet – needs to be worn in some activities (e.g. harvesting, bagging, crossing, and spraying for pests and diseases). A helmet protects against collisions with a range of objects (e.g. the operator falling, or falling palm fronds and bunches, and equipment).
- Safety eyeglasses – need to be worn to avoid splinters and splashes that can injure eyes. These should be worn when working with chemicals such as pesticides (when spraying for pests and diseases), fungicides (when bagging) and pollen, which may injure the eyes (when crossing).
- Safety respirator mask – this should be worn when carrying out activities that involve chemical materials or any hazardous substance that can be inhaled, such as pesticides when spraying, and pollen when crossing.
- Rope and safety harness – these should be used when bagging and crossing to avoid falling when working in palm trees. This is particularly dangerous when working in tall trees. Rope and safety harness have to be used when working on a palm over 3 m tall.
- Apron – to be worn when preparing chemical sprays (e.g. for pest and disease control).

Health and safety training and refresher training are essential for worker safety. The International Labour Organization (ILO) and the Round-table on Sustainable Palm Oil (RSPO) are international organisations that have concerns for worker safety. They provide guidelines and additional considerations about health and safety covering the following (see for example ILO, 2001; RSPO, 2013).

- checking that equipment works properly before use
- chemical use (especially pesticides)
- standard operating procedures (SOPs)
- working alone
- emergency procedures, first-aid box
- being aware of nuisance insects (e.g. mosquitoes) and other nuisance and dangerous animals (e.g. snakes).

2.2   Health and Safety in the Nursery

Although the work risk in the nursery is not as high as in the field, health and safety issues need to be recognised and observed.

Safety equipment, which needs to be worn in the nursery includes:

- hats and sun cream for sun protection
- safety boots
- gloves for handling sharp or hard materials
- masks if working with hazardous chemicals such as pesticides.

2.3   Health and Safety in the Laboratory

General guidelines for working at the laboratory bench are given by Barker (2005) and for safety in an analytical laboratory by Petrozzi (2013). Field trialling involves various laboratory procedures in analysing different performance traits. Different laboratories are used in field trialling work and although general health and safety procedures apply, each may have specific procedures. Bunch analysis of oil palm involves various hazardous laboratory procedures such as chopping the bunch stalks with an axe, scraping the mesocarp with a knife and oil extraction using hazardous solvents.

Greater rigour in field trialling is achieved with uniform planting materials. This is particularly important for fruit type (dura, tenera and pisifera). These types can be differentiated by laboratory DNA testing prior to field planting. Both bunch and oil analysis and DNA screening may be outsourced, but if these are done in-house then appropriate laboratory health and safety considerations need to be observed.

Basic laboratory safety issues include the following:

1. Identify hazards and implement appropriate safety control measures. Mark the work area with warning signs/stickers and contact information.
2. Do not eat, drink, smoke, apply cosmetics (including lip balm) nor handle contact lenses.
3. Do not store food or drink in laboratory refrigerators.
4. Do not wear open-toed shoes.
5. Do not allow visitors, including children, in laboratories where hazardous substances are stored or are in use, or where hazardous activities are in progress.
6. Protective clothing such as a laboratory coat (lab coat) must be worn before entering the laboratory building and removed when leaving the laboratory. The lab coat protects the body from the harmful effects of chemicals and provides protection to the samples being worked with; it also protects people outside the laboratory from contamination by laboratory materials.

Lab coats may be located in an airlock where shoes may also be changed for lab shoes and where hand-washing facilities may be located (for both entering and exiting the lab).

7. Use specific PPE during the handling of chemicals in the laboratory. Use goggles as eye protection, gloves as skin protection from heat and chemicals, and a mask as a respiratory protector.

8. Be aware of emergency procedures: building maps, evacuation routes, fire-fighting procedures, location of fire extinguishers, emergency signals, emergency shower, emergency exits, emergency personnel, emergency phone numbers, emergency meeting points, first aid-box/first-aider personnel and local medical facilities which are nearest to the laboratory building.

9. Be aware of hazards relating to the chemicals used in the laboratory and their safety data sheet (SDS). SDSs give full information about the chemical identification, hazard(s) identification, composition/information on ingredients, first-aid measures, fire-fighting measures, accidental release measures, handling and storage, exposure controls/personal protection, physical and chemical properties, stability and reactivity, toxicological information, ecological information, disposal considerations, transport information, regulatory information and other relevant information. The SDS should be printed out and placed in a designated place where it can be found easily.

10. Be careful when using sharp tools (e.g. using an axe to chop fruit bunches to separate the stalk and spikelets and using a knife to scrape the mesocarp from the fruit). Full training is required for these operations.

11. Be careful when working with glass, especially when handling glassware (e.g. beakers, boiling flasks and Soxhlets).

12. Be careful when using flammable solvents to prevent fire. Do not expose the solvent to an uncontrolled heat source or electrical plug.

13. Be aware of solvent air-exposure during solvent extraction operations. Solvent extraction should be carried out in a specialised room with air extraction systems above the soxhlets and with ventilation.

14. When using n-hexane during oil extraction refer to the SDS (Merck, 2016). Hexane is a dangerous chemical and needs to be handled with extreme care. Hazard statements of n-hexane:

- H225 – highly flammable liquid and vapour
- H304 – may be fatal if swallowed and enters airways
- H315 – causes skin irritation
- H336 – may cause drowsiness or dizziness
- H361 – suspected as damaging fertility; suspected of damaging the unborn child
- H373 – may cause damage to organs (nervous system) through prolonged or repeated exposure if inhaled
- H411– toxic to aquatic life with long-lasting effects.

Precautionary statements of n-hexane:

- Prevention:
  - P210 – keep away from heat, hot surfaces, sparks, open flames and other ignition sources; no smoking
  - P240 – ground/bond container and receiving equipment
  - P273 – avoid release to the environment.
- Response:
  - P301 + P330 + P331 – if swallowed: rinse mouth; do not induce vomiting
  - P302 + P352 – if skin contact: wash with plenty of soap and water
  - P314 – seek medical advice/attention if you feel unwell.
- Storage:
  - P403 + P233 – store in a well-ventilated place. Keep container tightly closed.
15. Safety consideration for laboratory DNA extraction and analysis.
    - Laboratory coats are mandatory in DNA laboratories. Wear gloves and safety goggles when handling hazardous material.
    - When leaving the laboratory, remove gloves and lab coats and wash your hands. Do not wear lab coats and gloves in offices, seminar rooms, toilets, kitchens, etc. Outside the laboratory area, all laboratory clothing (coats and gloves) must be removed.
    - Hands should be washed after handling biological materials, after removing gloves, and when leaving and entering the laboratory.
    - Safe use of centrifuges. The location of centrifuges is a major health and safety issue.
      - Before use – check the over-fill, balance, caps or stoppers are correctly placed.
      - Use sealable buckets (safety cups) or sealed rotors.
      - After the centrifuge run, make sure that the centrifuge has stopped completely and check for spills or leaks in the centrifuge chamber.
    - Do not pipette by mouth; only mechanical pipetting devices should be permitted.
    - Perform all procedures carefully to minimise splashes and aerosol sprays.
    - The laboratory should be kept clean and free of non-essential equipment pertinent to the work; work surfaces should be decontaminated at least once a day with 70% alcohol.
    - Any spillage must be cleaned up immediately. Put up warning signs if the floor is wet.
    - The use of needles and syringes should be restricted. When used, care must be taken to avoid injuries. Dispose of needles, syringes, broken glassware and other sharp objects into appropriate containers.

- Contaminated liquids or solid materials must be decontaminated before disposal.
- Use bottle carriers, carts or other secondary containers when transporting chemicals in breakable containers (especially 250 ml or more) through corridors or between buildings (e.g. from a chemical repository to the laboratory, or from the laboratory to a waste disposal facility). The individual transporting the chemical should know the hazards of the chemical and should know how to respond to spillage.
- Know the location and how to use emergency equipment, including safety showers, eyewash stations and medical cabinets.
- Be alert to unsafe conditions/actions. Report them to your supervisor immediately.
- For intermediate storage, waste must be collected in designated containers and places, and labelled 'solid biological waste' or 'liquid biological waste'.
- Solid waste: gels or any solid material that has been in contact with cells (e.g. Petri dishes, pipette tips, toothpicks, cuvettes, soaked paper towels, etc.). Solid waste may be autoclaved in special biohazard bags at 121°C for 20 mins.
- Non-hazardous liquid waste: before disposal in the drain, cultures and supernatants have to be either disinfected with sodium hypochlorite (bleach) solution (1–5% for at least 10 minutes) or equivalent (e.g. Sanosil© according to manual), or autoclaved at 121°C for 20 minutes.
- Many lab accidents happen due to inappropriate/insufficient protective clothing. If the worker is handling strong corrosives (NaOH, KOH, acids, phenol, HCl, phosphoric/acetic acid) or liquid nitrogen/propane/ethane, make sure the worker wears protective clothing and safety goggles. Some common hazardous chemicals used in DNA laboratories are listed here:
  ○ acrylamide: toxic, neurotoxic, possibly carcinogenic, harmful to environment
  ○ cycloheximide: very toxic, harmful to environment
  ○ hygromycin: very toxic, harmful to environment
  ○ ampicillin: noxious, harmful to environment
  ○ ethidium bromide: very toxic, carcinogenic, harmful to environment
  ○ Hg compounds: very toxic, harmful to environment
  ○ liquid nitrogen: can cause strong cold burns upon prolonged contact with skin; avoid splashes to the eye and soaking of clothes (e.g. gloves and lab coats)
  ○ liquid ethane: causes immediate strong cold burns on skin
  ○ NaOH, KOH, HCl, phosphoric/acetic acid: corrosive

- ○ Ni, Co compounds: toxic, allergenic, harmful to environment
- ○ phenol: toxic, corrosive
- ○ uranyl acetate: very toxic, radioactive, carcinogenic, harmful to environment
- ○ sulfuric acid: toxic, corrosive, harmful to environment
- ○ sodium azide: very toxic, harmful to environment.
- • Before handling hazardous substances, always read the material safety data sheets (MSDS) and follow the instructions specified therein.
- • Pay attention when handling hazardous substances (corrosives, toxics, flammables, etc.). Always work in a chemical fume hood and wear protective clothing and equipment (gloves, safety glasses, laboratory coat, shields, etc.).

16. Be aware of SOPs that have been developed for the laboratory, or which should be developed (e.g. maximum capacity of chemical storage, storage procedures, waste handling, waste disposal, etc.).

17. Maintain a laboratory notebook.

References

Barker, K. (2005) *At the Bench: A Laboratory Navigator*. Cold Spring Harbor Press, New York, NY, USA.

IFC (2016) *Environmental, Health, and Safety Guidelines: Perennial Crop Production*. Available at: http://documents.worldbank.org/curated/en/998041479462875256/pdf/110345-WP-FINAL-Perennial-Crop-Production-November-2015-PUBLIC.pdf, accessed 27 September 2019.

ILO (2001) *Convention 184: Convention Concerning Safety and Health in Agriculture*. Available at: https://www.ilo.org/public/english/standards/relm/ilc/ilc89/pdf/c184.pdf, accessed 25 July 2018.

Merck (Millipore team) (2016) Safety data sheet: 104374_SDS_Hexane EN. Revision 31.10.2016, version 4.4, Germany, Merck.

Petrozzi, S. (2013) *Practical Instrumental Analysis: Methods, Quality Assurance and Laboratory Management*, 1st edn. Wiley-VCH Verlag GmbH & Co, KGaA. Germany.

RSPO (2013) *Principles and Criteria for the Production of Sustainable Palm Oil*. Available at: https://rspo.org/publications/download/4b4296c7bb85cb3, accessed 25 July 2018.

Pre-trialling Considerations and Activities

3

Abstract

Before trialling can begin, various considerations and preparations need to be made. First it is beneficial to know if the trait of interest is governed mainly by genetic or environmental factors. If the trait is controlled by genes which are little affected by the environment then field trials may be set up in a narrow range of environments. If, however, the trait is strongly affected by environment, for example, soil type or climate, then trials need to be more extensive to give an estimate of the environmental and genotype x environment interaction. Multi-locational trials can determine the extent of genetic versus environmental factors involved. Plant breeding is an essential pre-trialling activity as it supplies progeny that have potential for crop improvement (see Setiawati *et al.*, 2018, this series). Field performance testing, in turn, is essential in selecting the progenies or palms sought. Materials for field trialling are seedlings. These are made ready for trialling in the nursery (see Laksono *et al.*, 2019, this series). Coordination is therefore required in preparing planting materials and land preparation for the trial. These aspects are discussed.

3.1 Genetic versus Environmental Effects

Genotype (G), environment (E) and genotype by environment (GxE) effects occur depending on the level of response of the genotype to various environmental factors and how different genotypes vary in their responses. A simple example: two palms may have a similar yield potential, but one is susceptible to disease and the other is resistant. In the absence of disease, both yield the same, but when the disease is prevalent, the resistant palm outyields the susceptible palm.

The observed variation (phenotype) reflects both the inherent genetic and environmental effects and the interaction of the two. Palms with stable yields under a wide range of environments are of great interest to breeders and commercial growers.

Oil palm is a perennial crop, which once established is productive for many years (about 20). It is therefore expected that an oil palm plantation will experience a range of biotic and abiotic stresses during its life span. Thus, developing and exploiting GxE effects to maximise yield in different environments is beneficial to oil palm growers (Corley and Tinker, 2015).

3.2   Need for Trials

Planters of any crop will use the best available materials. Growers may hear that other planters/farmers had success with a particular variety, but without comprehensive trialling it cannot be proven until they grow it themselves. Comparisons cannot be made without good planning. For example, it is not a simple matter of planting two varieties/genotypes at the same location and measuring performance, as we cannot say that one is better than the other as there is no robust way to compare the two without replication. Furthermore, it is preferable to test more than two varieties/genotypes as breeders can produce several progenies.

Most trial work tries to verify the hypothesis that one variety/genotype/line is better than the others. In other instances, theoretical considerations may play a major role in arriving at a discriminative test hypothesis. For example, it can be shown theoretically that a certain genotype removes more nitrogen from the soil than is naturally replenished. Thus, we may hypothesise that to maintain the yield potential, supplementary fertiliser must be added. Once the hypothesis is framed, the next step is to design a procedure for its verification. This experimental procedure usually consists of four parts.

1. Selecting the appropriate material to test.
2. Specifying the characteristics to measure.
3. Selecting the procedure to measure these characteristics.
4. Specifying the procedure to determine whether the measurements made support the hypothesis.

The first two are a relatively easy matter for the oil palm breeder as the traits and available germplasm are normally well known. For example, in oil palm, the test material would probably include newly developed advanced lines and standard contemporary varieties (for comparison). The characteristic to be measured would probably be yield (Chapter 7). The other two parts (3 and 4 above) will be discussed in Chapters 6 and 7.

The main breeding objectives are as follows:

1. High yield (Chapter 7, 1 and 2).
2. Disease resistance/tolerance, especially *Ganoderma* (Chapter 8).

3. Adaptability – good performance over a wide range of environments.
4. Easy harvesting (refer to novel traits, Chapter 7.6).
5. Bunch character (high oil to bunch, big kernel, Chapter 7.5).
6. Low vegetative vigour.
7. Pest tolerance.

3.3 Breeding

A broad genetic base is needed in order to capture and exploit genes to produce new superior varieties. Oil palm breeding companies and research stations have, and continue to expand, germplasm collections from native sources in Africa (Nigeria, Cameroon, Angola, Ghana, Tanzania etc., Chapter 1). Broadening the genetic base of breeding material will help breeders to capture a wide range of genetic variation and to start introgression into elite breeding materials.

Dura, Pisifera and Tenera

Breeding in oil palm started in Nigeria with very limited germplasm, initially confined to dura (D) and used in DxD crosses both for breeding and for commercial seed production. Later the use of pisifera palms as male parents crossed onto dura females was found to be more interesting and produced higher yielding tenera types for commercial seed production.

Deli dura
Although it is believed that all Deli duras are descendants of only four palms (founding palms planted in the Botanic Gardens, Bogor, Java, Indonesia in 1848), the population expanded to several million before systematic breeding programmes began. These breeding programmes often overlapped, even though they were carried out independently in several locations and in different countries – for example, at Gunung Bayu, Pabatu, Dolok Sinumbah, Marihat Baris, Mopoli/Bangun Bandar and Gunung Melayu in Indonesia, at Serdang Avenue, Elmina, Ulu Remis and Johor Labis in Malaysia and at Dabou in Côte d'Ivore (Rosenquist, 1985) and Papua New Guinea (Sterling and Alvarado, 2002).

Other dura populations originate from the following groups (see Rosenquist, 1985 for more details).

- *Angola*. The main advantage offered by the Angola population (when used as a female progenitor) is reduced stem growth rate; however, bunch production is less than in Deli dura lines. Angola mother genotypes combine well with Ekona and Mardi as male parents.

- *Bamenda*. Although not extensively tested, the Bamenda population exhibits potential for large fruit bunches with reduced palm stem growth. Annual bunch yields for Bamenda x AVROS (Algemeene Vereniging van Rubber-planters ter Oostkust van Sumatra) progenies are about 200 kg plant^{-1} yr^{-1} with a reduced stem growth and acceptable bunch composition. This results in a high commercial potential of over a ton of oil per ha per year over standard controls.
- *Kigoma*. DxP progenies obtained from Kigoma female progenitors crossed with several male parent sources showed a high bunch and oil yield potential, very similar to tenera materials derived from Deli duras. This is particularly the case when the male source is from Mardi or AVROS origins. Vegetative growth of Kigoma genotypes is relatively low.
- *Other uncommon sources of dura material*. An experiment planted in 1990 in Santo Domingo de los Colorados, Ecuador, showed that the Deli and Deli x Angola female lines, and the Ekona male line were very precocious with high general combining ability (GCA), while the best specific combinations were Deli x Yangambi and Kigoma x Ekona progenies.

Pisifera

Pisifera palms are normally female sterile and this precludes any field trial evaluation for yield. The only direct measurements that can be obtained from pisifera plants are morphometric and leaf mineral content (fertiliser-use efficiency). The most commonly recorded traits are leaf area index (LAI), stem growth rate, leaf length, leaf emission rate and magnesium content. The strategy for breeders is therefore to evaluate the performance and production of the tenera siblings arising from the same family or, even better, the performance of the DxP progenies directly.

Thus, selection based on progeny testing appears to be the only reliable criterion to decide the genetic potential of a pisifera genotype, with the aim of selecting male parents for commercial oil palm seed production. As an example: ASD has been progeny testing its original introduced D'jongo derived sources of male parents since 1969 (Sterling and Alvarado, 2002).

Tenera

Since pisifera genotypes are often female sterile, developing the pisifera to produced improved pisifera material is very difficult. Pisifera improvement via breeding through tenera population production is currently the best way to produce new generations of pisiferas: selfing teneras will produce segregation of all fruit types – dura, tenera and pisifera – with an expected 1D:2T:1P ratio.

The most famous TxT progeny was that which produced the tenera genotype SP540 at Sungai Pantjur. This palm was selfed to produce the progeny Pol 820 in 1931. In 1973 Research Institute of the Sumatra Planters Association (RISPA) repeated the selfing of SP540 and produce another generation from three progenies, all of which were planted at Aek Pantjur, Indonesia, in 1973. The pisifera descendants went on to be highly successful male parents in commercial seeds production (Rosenquist, 1985).

Germplasm acquisition and collections

The need for more extensive germplasm collections to broaden the genetic base of the oil palm breeding material, particularly in Asia, and to safeguard future cropping, has become imperative. With respect to other major crops, oil palm collections are relatively recent as oil palm commercialisation did not begin until the early 20th century.

The first organisation to start collecting oil palm from its natural environment was WAIFOR (West African Institute for Oil Palm Research), Nigeria (described in Forster *et al*. 2018). This began in 1911. The headquarters of the Institute is located at its main station in Benin City. The research station established a sub-station at Abak in the Calabar province (182 ha). Other agricultural stations utilised in the collection programme were: Moor Plantation (Ibadan); Ogba (Benin); Nkwele (Onitsha); Umudike (Umahia); and Obio-Akpa (Abak). Unfortunately, this programme was interrupted by war and was not re-started until 1952.

However, collections were made in Calabar, Aba, Ufuma, Angola from 1911. Breeding programmes were then initiated to exploit the material. In 1926, WAIFOR received Deli material from SOCFIN-Tanjung Gentung and AVROS from Indonesia. Later, Deli genotypes were received from Serdang Centre Experimental Station, Malaysia and IRHO (CIRAD) (Richardson and Alvarado, 2003). Recent oil palm collections were carried out in the highland area of Afikpo in Ebonyi State (Okwuagwu *et al*., 2011).

In 1973, Malaysia, under the auspices of PORIM (currently known as Malaysian Palm Oil Board – MPOB) began collecting wild material from Africa as it was acknowledged that the gene pool of planting materials was very limited. Most African countries where oil palm can be found were explored, including Nigeria, Cameroon, DR Congo, Tanzania, Madagascar, Angola, Senegal, Gambia, Sierra Leone, Guinea and Ghana (Rajanaidu and Jalani, 1994).

Collections were located at the Ghana Oil Palm Research Station as this was part of WAIFOR at the time. Breeding programmes were subsequently started in Ghana using these materials. Later, in 2004, wild germplasm collections were initiated in other regions of Ghana. Prospection in the northern regions included the areas of: Vogoni Korri Nadowli, Vogoni Korri

Forest Nadowli, Bugri Corner Koka Bawku, Saaka Bawku and Damango Canteen Bredi Farm (Sapey *et al.*, 2012).

Collaboration among oil palm research institutes in Indonesia to collect, conserve and exploit germplasm from Nigeria has been established.

3.4   Preparation of Materials for Trialling

Breeding trials are carried out on progenies from deliberate matings. Fruits are collected and seeds are processed (see Setiawati *et al.*, 2018 for crossing methods and Kelanaputra *et al.*, 2018 for seed production methods, this series). The germinated seeds are then reared in the nursery to obtain field-ready plants (see Laksono *et al.*, 2019, this series).

Pre-planting screening is now a possibility using DNA analyses. Nursery seedlings offer a convenient phase in which to sample leaves, extract and analyse DNA. Genes or diagnostic DNA markers for characteristics of interest can be screened in the laboratory – for example, fruit type (the classification of dura, tenera and pisifera genotypes), fruit colour (virescent or nigrescent) and mantled fruit. Screening prior to field trial planting is especially important for fruit type as the sterile pisiferas can be eliminated and 'dura only' and 'tenera only' plots can be set up, thus providing greater rigour to the trialling and more reliable and relevant results obtained.

3.5   Land Selection and Preparation for Trialling

Aerial land surveys using drone-captured images are becoming more common in selection of land for trialling. Aerial images provide information on factors that need to be taken into consideration such as topography, waterways, existing crop/use, roads, supply lines, buildings and conservation areas. The information may also be used in designing the trial (Fig. 3.1). In the case of *Ganoderma* trials it can also determine the extent of disease prior to planting.

There are principally two sources of land used for trialling, based on the previous land use, for example ex-oil palm plantations and 'ex-other' plantations (e.g. rubber). Trialling in primary and secondary forest is not sustainable and should definitely be avoided. Methods of land preparation start with felling the previous crop, then clearing the area and preparing the trial site. Procedures in land preparation should consider national and international regulations, which are often requirements in sustainable and eco-friendly practices. Preparation of land for trialling will be discussed further in Chapter 4.

Pre-trialling activities in oil palm breeding programmes prior to land preparation are shown in Table 3.1.

Fig. 3.1. Aerial image of a trial site overlaid by a trial design.

Table 3.1. Pre-trialling activities in oil palm breeding programmes prior to land preparation.

Activity	Year 1	Year 2	Year 3	Year 4
Identify target traits	▓			
Assess germplasm for required variation	▓			
Acquire variation (if needed)	▓			
Have parental materials (mother palms and pollen) ready	▓			
Carry out crossing		▓		
Harvest bunches		▓		
Process seed		▓		
Grow seedlings in nursery			▓	
Transfer young plants to field for trialling				▓

References

Corley, R.H.V. and Tinker, P.B. (2015) *The Oil Palm*, 5th edn. World Agriculture Series. Wiley Blackwell, Hoboken, NJ, p. 639.

Forster, B.P., Sitepu, B., Setiawati, U., Kelanaputra, E.S., Nur, F. *et al.* (2018) Oil palm breeding. Campos, H.A. and Caligari, P.D.S. (eds). *Genetic Improvement of Tropical Species*. Springer, New York, pp. 241–290.

Kelanaputra, E.S., Nelson, S.P.C., Setiawati, U., Sitepu, B., Nur, F. *et al.* (2018) *Seed Production in Oil Palm: A Manual. Techniques in Plantation Science*. Forster, B.P. and Caligari, P.D.S. (eds). CAB International, Wallingford, UK, p. 59.

Laksono, N.D., Setiawati, U., Sitepu, B., Nur, F., Forster, B.P. *et al.* (2019) *Nursery Practices in Oil Palm: A Manual. Techniques in Plantation Science*. Forster, B.P. and Caligari, P.D.S. (eds). CAB International, Wallingford, UK.

Okwuagwu, C.O., Ataga, C.D., Okoye, M.N. and Okolo, E.C. (2011) Germplasm Collection of Highland Palms of Afikpo in Eastern Nigeria. *Bayero Journal of Pure and Applied Sciences*, 4: 112–114.

Rajanaidu, N. and Jalani, B.S. (1994) Oil palm genetic resources collection, evaluation, utilization and conservation. Paper presented at *PORIM Colloquium on Oil Palm Genetic Resources*. PORIM, Bangi, Malaysia.

Richardson, D. and Alvarado, A. (2003) ASD Oil Palm Germplasm from Nigeria. *ASD Oil Palm Paper* 26: 23–32.

Rosenquist, E.A. (1985) The Genetic Base of Oil Palm Breeding Populations. *International Workshop on Oil Palm Germplasm and Utilisation*, Malayasia.

Sapey, E., Adusei-Fosu, K., Agyei-Dwarko, D. and Okyere-Boateng, G. (2012) Collection of oil palm (*Elaeis guinensis* Jacq) germplasm in the northern region of Ghana. *Asian Journal of Agricultural Sciences*, 4: 325–328.

Setiawati, U., Sitepu, B., Nur, F., Forster, B.P. and Dery, S. (2018) *Crossing in Oil Palm: A Manual. Techniques in Plantation Science*. Forster, B.P. and Caligari, P.D.S. (eds). CAB International, Wallingford, UK, p. 50.

Sterling, F. and Alvarado, A. (2002) Historical account of ASD's oil palm germplasm collection. *ASD Oil Palm Papers* 24: 1–16.

Land Preparation

4

Abstract

Previous land use determines the operations needed in preparing land for trialling. Trialling in primary and secondary forest is not a sustainable nor an environmentally friendly practice and should not be carried out. In sustainable oil palm production, previous land should normally be a previous plantation crop (e.g. rubber, coconut or oil palm). The procedures for the felling of two of these plantation crops, clearing the land and preparing the trial site ready for planting, are given as examples here. These procedures need to be carefully timed and coordinated with nursery plant production and season (oil palm is normally planted in the rainy season). They are also regulated by the need for and involvement of eco-friendly practices – for example, 'zero burning, zero smoke', in clearing the land area. Heavy specialised equipment is used throughout and operators and field work must observe standard operating procedures.

4.1 Ex-rubber Preparation

Felling rubber trees

Felling rubber provides additional revenue as the timber can be sold for wood – in addition, with previous planning, the rubber can be 'slaughter tapped' to gain maximum latex production before the trees are felled. The contract of selling rubber wood is normally agreed before the work begins. All the felling and clearance work may be done by a contractor, but the contractor must follow all standard operating procedures (SOPs) stated in company regulations. All operators/workers must obey the standard health and safety procedures involved (Chapter 2).

The land area (block) selected for felling is marked at the corners to demarcate it and prevent/minimise any misunderstandings (e.g. the wrong block being felled, or parts of other blocks being felled). The felling process is often used to train and provide refresher training in operations and is supervised throughout to prevent accidents (from felled trees, misuse of chainsaws and movement of field vehicles). The operator cuts the rubber trees with a chainsaw (Fig. 4.1) as low as they can and, of course, in the opposite direction to the one in which the tree will fall. Branches of the felled tree are removed by cutting with a chainsaw, the rubber wood is cut into pieces of about 2 m, uploaded into a truck for removal, and transferred to a wood factory by the contractor.

After felling and clearing are completed, the rubber tree roots are unearthed using an excavator. If the area is intended for a seed garden trial, all the rubber stumps must be collected and removed from the area (to minimise risk of residual *Ganoderma*, a major soil-borne disease of oil palm (Cooper *et al.*, 2011)), while for other trials the rubber roots are normally stacked in the 'deadline row' (the row that is used as a harvesting road or non-collective road). Stacking is normally done at four oil palm row intervals.

Fig. 4.1. Felling a rubber tree with a chainsaw.

Fig. 4.2. All rubber roots and small branches are allowed to dry, and then removed and gathered outside the plantation.

After uprooting, all the rubber roots and branches are left in the field (Fig. 4.2) to dry for one to two weeks and then removed from the block by the contractor and placed into designated areas to decompose. This procedure is part of an environmentally friendly zero-burning system.

Lining

After the removal of all rubber roots and branches, the next step is to mark the positions of roads and drains for the trial site by lining. Drains are excavated (Fig. 4.3) where needed to allow flow of water and to prevent flooding, particularly in the rainy season. For trial purposes it is best if the land is uniform to prevent excess variation in data due to localised environmental factors; if there is limited land, the blocking should follow the drain line, therefore the drain must be as straight as possible (to prevent the drain crossing plots in the trial). Likewise, in terraced areas the drain should follow the slope contour as much as possible.

Ripping

The objective of ripping is to clean the soil from the roots of rubber trees which may still remain in the earth, as this a potential source of

Fig. 4.3. Excavation of a drain with heavy machinery.

Fig. 4.4. Scheme for ripping directions to remove roots (left) and heavy machinery with ripper attachment (right).

oil palm diseases. The ripping work is carried out four times (Fig. 4.4), using a tractor equipped with a ripper attachment, with an operational depth of 60 cm.

A ripper is normally deployed diagonally (Fig 4.4) to the ex-rubber block (north-east to south-west) with a second ripping at right-angles to unearth all the roots from the soil. After rippings 1 and 2, all rubber roots are collected and placed at the roadside. The work is repeated for rippings 3 and 4, thus different ripping directions are achieved.

Ploughing

The objective of ploughing is to turn over the soil to expose any remaining rubber roots. Ploughing is carried out using a tractor with equipment operating at a depth of 30 cm. Because large, heavy equipment is used to clear the site and to do the ripping, the soil will be compacted and should be loosened and aerated. Ploughing is carried out twice (Fig. 4.5): the first ploughing direction is in a diagonal direction, from north-west to south-east, the second from north-east to south-west. After ploughing, any unearthed roots are collected and moved to the roadside. This is carried out for areas totally cleaned of rubber material, while for areas where rubber roots have been stacked in the block, ploughing is carried out from north to south and vice versa.

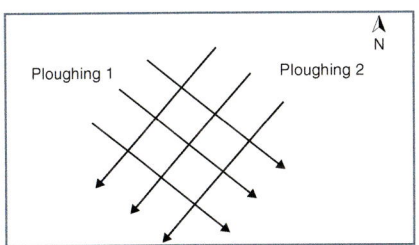
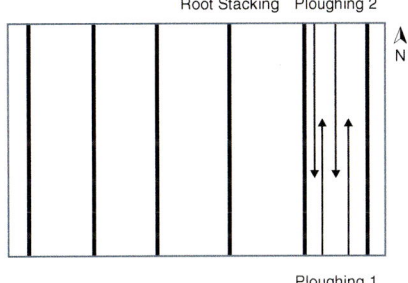

Fig. 4.5. Ploughing scheme for areas cleaned of rubber roots (left) and for areas where roots have been stacked in the field (right). Ploughing equipment mounted to a tractor can be seen in the photograph.

Harrowing

Harrowing is conducted after ploughing (Fig. 4.6). As for ploughing, harrowing is carried out twice where the first harrowing is diagonal, from north-east to south-west, and the second runs north-west to south-east. After the second harrowing, any unearthed roots are again collected and placed on the roadside. This process is carried out in areas totally clear of rubber materials, but for areas where rubber roots were stacked, harrowing is carried out from north to south, then east to west.

Excavation of main drain and collecting drains

The main drain and collecting drains are constructed to provide water flow in the planting area. Water should not gather in the trial area. Collecting drains are normally made in deadline rows.

Excavation of collection road

The collection road allows workers to access and harvest the trial area. Collection roads play an important role as they help to keep the fruit fresh

Fig. 4.6. Tractor equipped with harrow attachment.

and with minimal damage (fruit can be broken by excessive bumping when bunches are moved from palms to the harvesting collection point). Collection roads are also useful during transport of seedling palms from the nursery to the field.

Spraying to control weeds

Any weeds are best removed and destroyed because many are hosts to potential pests and diseases. Weeding is preferably avoided or carried out by hand, but where necessary it can be performed by herbicide application, using the least harmful herbicide possible.

Mucuna planting

Mucuna spp. are legume cover crops (LCCs), which fix nitrogen (Mathews, 1999). These plants are very vigorous and help to cover the soil to reduce weed growth, erosion and run-off. *Mucuna* spp. are best planted before oil palm planting but can be planted simultaneously.

Table 4.1. shows land preparation for oil palm ex-rubber in North Sumatra, Indonesia, prior to the planting season which starts in September and ends in February (the rainy season).

4.2 Ex-oil Palm Preparation

Oil palm companies normally have a set of defined environmental policies – for example, in accordance with the principles of RSPO (Round Table on Sustainable Palm Oil), where practices in land clearing, replanting and zero burning apply. Benefits include:

1. Prevent air pollution from smoke.
2. Maintaining soil aggregate and structure, as well as the soil ecosystem.
3. Prevent fires and carbon emissions released from peat lands.

Various sequences of land preparation have been followed in different parts of the world. The methods chosen are often determined by local circumstances and experience. In previous times it was usual to burn the underbrush (clear the ground vegetation and small saplings) and line out before felling (Hartley, 1988). This is no longer recommended nor is it standard practice.

Felling oil palms is divided into two practices based on the trial to be set up. For normal performance and progeny testing, the oil palms are felled

Table 4.1. Land preparation for oil palm ex-rubber.

No.	Activity	April	May	June	July	August	Sept.
1	Felling	▓					
2	Uprooting		▓				
3	Removal of roots and small stems from trial site		▓				
4	Lining for stacking, roads and drains		▓				
5	Stacking						
6	Ripping 1 and 2			▓			
7	Collecting roots 1			▓			
8	Ripping 3 and 4			▓			
9	Collecting roots 2				▓		
10	Ploughing 1 and 2				▓		
11	Collecting roots 3				▓		
12	Harrowing 1 and 2					▓	
13	Collecting roots 4					▓	
14	Excavate main drain 2 x 1 x 1 m						
15	Excavate collecting drain 1 x 1 x 0.75 m						
16	Excavating collection road 6 m wide			▓			
17	Removing pre-emerged weeds					▓	
18	Lining planting point						▓
19	*Mucuna* planting						▓

The "Month" heading spans the April–Sept. columns.

Note: if estate practices include a fallow period, then after planting *Mucuna* the land will remain fallow for up to a year before planting oil palm.

at least a year before the trial is to be planted. This is an SOP to prevent palm overlaps that promote pests and diseases. However, non-fallowing is useful when field trialling for *Ganoderma*. In this case preparations for trialling involve felling three months before planting because a major objective is to *encourage* the development of *Ganoderma* to allow screening for *Ganoderma* resistance. These trials should be as homogenous as possible to provide equal opportunities for *Ganoderma* infection in the trial.

Felling of oil palm trees is usually achieved using an excavator (Fig. 4.7). The line of the work must be drawn and a tall, visible pole (3 m high) is set up to allow the operator to see and maintain the felling direction. The oil palms are felled by pushing them over with an excavator (of course, ensuring the palm falls away from the excavator). After felling, the oil palm

Fig. 4.7. Tractor pushing over an oil palm tree.

trunk is removed to the path using an excavator and then removed again to a designated site or chipped in situ (Fig. 4.8).

Replanting oil palm has a high risk of pest and disease, especially if the replanting period is on second and third generation oil palm. *Ganoderma* (basal stem rot) is a particular concern in Southeast Asia because the disease becomes more prevalent with each replanting of oil palm.

Fig. 4.8. Heavy machinery used in chipping the oil palm trunk after felling.

Table 4.2. Land preparation for oil palm from ex-oil palm.

No.	Activity	Month		
		Aug.	Sept.	Oct.
1	Lining for stacking, roads and drains			
2	Felling			
3	Uprooting			
4	Stacking and removal			
5	Ploughing 1 and 2			
6	Harrowing 1 and 2			
7	Chipping the oil palm trunk			
8	Removing pre-emerged weeds			
9	Lining			
10	*Mucuna* planting			

Note: if estate practices include a fallow period, then after planting *Mucuna* the land will remain fallow for up to a year before planting oil palm.

Once the land has been cleared, the operations are similar to those described for ex-rubber land preparation (see Table 4.2).

Soil tillage for planting oil palm into ex-oil palm sites is the same as the soil tillage in ex-rubber. The difference, as seen in the table above, is that there is no ripping or root collecting work.

4.3 Equipment

Before land preparation work is carried out, training and socialisation of the planned work is provided to workers and people living near the project area and other interested parties. This must be carried out to minimise hazards and work accidents. SOPs must be followed as outlined in Chapter 2.

Tools and equipment used for land preparation are as follows:

1. Hand-held chainsaws are used to cut down rubber trees (in preparation of ex-rubber land). This must be done by trained personnel.
2. Glasses, helmets, boots and hand gloves are used to prevent and reduce accident risk (e.g. from falling branches).
3. Dumper trucks are used to remove wood and root debris from the field.
4. An excavator is used to push down oil palm trees and uproot rubber trees.
5. A tractor with appropriate attachments is used for ripping, harrowing and ploughing the soil.

References

Cooper, R.M., Flood, J. and Rees, R.W. (2011) *Ganoderma boninense* in oil palm plantations: current thinking on epidemiology, resistance, and pathology. *The Planter* 87(1024): 515–526.

Hartley, C.W.S. (1988) *The Oil Palm (Elaeis guinensis* Jacq.), 3rd edn, Longman Science and Technical, p. 361.

Mathews. C. (1999) The introduction and establishment of a new leguminous cover crop, *Mucuna bracteata* under oil palm in Malaysia. *Proceedings of the 1999 PORIM International Palm Oil Congress*, pp. 610–616.

Material Preparation **5**

Abstract

The aim of oil palm trialling is to evaluate field performance among a range of genotypes and to select those of interest. A basic necessity is the provision of good quality planting material, usually produced from deliberate cross-pollinations. Details of methods in pre-trialling activities, i.e. in crossing, seed production and the production of field-ready planting materials, are given in Kelanaputra *et al.* (2018), Setiawati *et al.* (2018) and Laksono *et al.* (2019), respectively (this series of manuals). Thus, only brief overviews of these operations are given here.

5.1　Crossing

Normally a prerequisite for breeding and subsequent trialling is the crossing of parental lines that combine traits of interest, although seeds collected from elsewhere, from open-pollinated bunches, etc., can be trialled. Currently, the major traits of interest in oil palm are yield, disease resistance and quality. More recently, traits that may be exploited in developing mechanical harvesting in oil palm have become increasingly studied and new eco-friendly traits (e.g. production of biodiesel and bioplastics) are becoming fashionable.

Pollen for crossing is collected from male inflorescences. These are isolated in bags prior to flowering. Specialised isolation bags are used that contain a pollen-collecting bag and a window to observe development. Male inflorescences are harvested at the end of anthesis. The collected pollen is sieved, dried, vacuum-packed and stored ready for use.

Pollen viability is tested. Fresh pollen has the highest viability, and viability declines with time in storage. Pollen is normally mixed with talcum powder to maximise the use of pollen and the ratio of pollen to talc is determined by pollen viability (the higher the viability the more the pollen can be diluted by the talc).

The developing female inflorescence is normally treated (using standard operating procedures, SOPs) with fungicides and insecticides to prevent disease. The inflorescence is covered with an isolation bag which is equipped with an inspection window, also used for spraying and pollination. The isolation bag provides insect-proofing, thus preventing uncontrolled pollinations.

Before pollination, the outside of the isolation bag is sprayed with an insecticide to kill any insects that may carry unwanted pollen (oil palm is naturally pollinated by insects). The pollen/talc mixture is blown onto receptive female inflorescences through the bag window. Labels showing pollination dates and parental genotypes are then attached and the bag is removed after fruit-set. The fruit bunch is harvested at maturity. The ripe bunches with red-coloured fruit (for nigrescent types) or orange-coloured fruit (for virescent types) are harvested about 150 days after pollination, or when between one and five loose fruits drop to the ground. Harvested bunches are sent, along with their labels, to the processing area in gunny sacks, along with any loose fruit that detach during harvesting. For details on practical procedures in oil palm crossing see Setiawati *et al.* (2018).

5.2   Seed Germination

In order to collect fresh seeds from harvested bunches, the fresh bunch is chopped to separate the spikelets from the stalk. Fruits attached to spikelets are fermented for three to four days to aid fruit abscission. The detached fruits are then peeled and placed in a de-pulping machine to remove the mesocarp from the seed (a nut containing a kernel). Any remaining fibres on the seeds are then scraped off. The seeds are disinfected and dried.

- Embryo viability testing is carried out to forecast germination rates.
- Seed moisture content and temperature are critical, to achieve the highest germination rates and both are carefully controlled.
- Oil palm seeds are naturally dormant. Dormancy is broken by seed processing and synchronised germination is achieved by heat treatments.

In commercial seed production, all seed has a thick shell (from the maternal dura parent). However, for breeding purposes tenera females are often used, and these produce thin-shelled seeds. Normally, tenera seeds have a lower germination rate compared to duras, but the aim is to produce sufficient germinated seeds for trial purposes. For practical details in oil palm seed germination methods see Kelanaputra *et al.* (2018).

5.3   Nursery

The performance of palms in trials is determined from an early stage by the quality of planting materials, which are normally seedlings, but ramets (usually produced from tissue culture cloning) are also used. Raising seedlings and ramets is done in a nursery. The aim is to produce uniformly healthy and vigorous plants for field trialling. Seedlings and ramets are needed for trialling (young plant screening and mature palm field performance testing) for a number of purposes: to assess progenies and breeding lines for selection and breeding; to assess materials for pest and disease resistance or tolerance to abiotic stresses, and responses to, and suitability for, changing agronomic practices and new planting materials.

There are normally two areas in a nursery, a pre-nursery (for germin-ated seeds in small bags or planting trays) and a main nursery (to grow on young plants in larger bags or planters). Abnormal palms are observed and discarded.

After two to three months in the pre-nursery, the palms are potted-on into large planting bags or planters in the main nursery. At this stage the seedlings (or ramets) are given more space.

Other activities in the nursery include fertiliser application, watering (irrigation), weeding and disease and pest control. Practical details of these activities along with others (e.g. how to set up an oil palm nursery) are given in Laksono *et al.* (2019).

New activities associated with the nursery now include DNA screen-ing to allow the selection of plants with specific traits, prior to potting-on and field planting. Thus, unwanted genotypes can be removed before field planting with savings in space and costs. This is particularly important in screening for shell type in distinguishing dura, pisifera and tenera palms. Thus, sterile pisiferas may be eliminated and trials can focus on dura and/ or tenera types. DNA testing is a rapidly developing area and, in the future, more traits will be selected by DNA testing in the nursery (Chapter 9).

References

Kelanaputra, E.S., Nelson, S.P.C., Setiawati, U., Sitepu, B., Nur, F., Forster, B.P. and Purba, A.R. (2018) *Seed Production in Oil Palm: A Manual. Techniques in Plantation Science.* Forster, B.P. and Caligari, P.D.S. (eds). CAB International, Wallingford, UK, p. 59.

Laksono, N.D., Setiawati, U., Nur, F., Rahmaningsih, M., Anwar, Y. *et al.* (2019) *Nursery Practices in Oil palm: A Manual. Techniques in Plantation Science.* Forster, B.P. and Caligari, P.D.S. (eds) CAB International, Wallingford, UK.

Setiawati, U., Sitepu, B., Nur, F., Forster, B.P. and Dery, S. (2018) *Crossing in Oil Palm: A Manual. Techniques in Plantation Science.* Forster, B.P. and Caligari, P.D.S. (eds). CAB International, Wallingford, UK, p. 96.

Trial Planting

6

Abstract

A brief introduction to selection is given. This is based on a natural process, but breeders aim to accelerate it by screening selected palms within and between genotypes or progenies under field conditions. Breeding is only effective if the trait of interest is controlled genetically (traits that are strongly influenced by the environment are better controlled by agronomy and husbandry). Screening is traditionally done by phenotypic scoring of plant performance in the field (e.g. by measuring yield), but, increasingly, traits may be selected for by DNA screening (genotypic selection). The advent of DNA screening now means that an increasing number of traits can be screened for at the seedling stage and selections can be made in the nursery (e.g. for shell thickness). Thus, field trialling can be more focused (e.g. tenera-only trials) and material of no interest eliminated (e.g. sterile pisiferas). DNA analyses can also aid in cross-selection (e.g. in conservative breeding – elite x elite crosses – or more speculative breeding – elite x wild, or wild x wild crosses), and provide a balance between the two. Once a breeding programme has been developed it needs to be coordinated so that material (from various crosses) is available for seed processing, nursery plant production and simultaneous field planting. In anticipation of this, a trial design needs to be considered (finalised when the material is realised) so that field preparations (appropriate land designated, clearance of previous crops, fallow, ploughing, plotting, drainage, etc.) and a field layout implemented. Various trial designs are discussed, and recommendations made depending on trial objectives and available materials.

6.1 Material Selection

Natural selection (i.e. evolution) has resulted in diversity of plant and animal life on earth. All selection results in a change of gene frequencies.

Throughout evolution, species have been changing, and better adapted genotypes become predominant, while those that are less fit move towards extinction. The aim of plant breeding is to mimic and accelerate selection of crop plants best suited for agricultural production and consumer needs. In order to be successful in a selection programme, three key criteria must be satisfied:

- variation exists within germplasm available to the breeder
- the desired trait(s) can be distinguished and detected phenotypically and/or genotypically
- the desired traits are under genetic control (Brown *et al.*, 2014).

Traditional breeding has been based on phenotypic selection, i.e. the observation of traits, usually in the field (e.g. yield, quality, presence of pest and diseases, flowering time, height, colour, etc.). In modern breeding, genetic variation can be detected by DNA analysis. DNA analysis can be used for various purposes (e.g. in marker-assisted selection, maintenance of elite genetic background, legitimacy testing and diversity analysis). These can be exploited in making big jumps forward in plant breeding and are now being applied to oil palm improvement.

Marker-assisted selection (MAS) and DNA genetic profiling of genetic stocks are both acknowledged to be powerful tools in plant breeding and have great potential for improving oil palm breeding. Several major oil palm breeding problems can be addressed by MAS and genetic profiling. A major trait of interest in oil palm is shell thickness. This is controlled by a single gene, *Sh* which in a homozygous state (*Sh/Sh*) gives thick-shelled fruits and, alternatively, when in the homozygous *sh/sh* state gives a pisifera, shell-less phenotype. The heterozygote, *Sh/sh* gives the commercial thin-shelled tenera form. DNA analysis can differentiate between these types at the seedling stage, i.e. before field planting (see Laksono *et al.*, 2019, this series).

DNA analysis for genetic diversity is another major application as germplasm collections can be screened efficiently for duplication/redundancy and thus field planting can maximise diversity or target specific genotypes of interest. Wening *et al.* (2009), for example, showed how genotyping could separate commercial from non-commercial and wild genotypes of oil palm.

Using DNA diversity analysis, the breeder can decide on a crossing programme which is expected to have a high possibility of producing better genetic combinations for traits of interest (e.g. yield). These may be among progeny of commercial x commercial crosses or more speculatively between commercial x non-commercial or even non-commercial x non-commercial crosses.

Other traits may also be considered before selecting material for field planting. *Ganoderma* resistance is one such trait. Currently there is no DNA test for *Ganoderma* resistance/tolerance, but this may be assessed phenotypically using a seedling screening test (see Rahmaningsih *et al.*, 2018, this series). However, the acid test for *Ganoderma* resistance, which is a major breeding goal, is by field trialling in a *Ganoderma*-endemic area. For this, a special trial needs to be set up in an ex-oil palm diseased area with no fallow period, as the aim is to have a high disease pressure present in the soil (see Chapter 8) on *Ganoderma* field trialling).

Typically, for a large-scale crossing design it is necessary to complete several hundred crosses. Although seed may be stored, it is better to use fresh seed. As germinated seeds should be planted at about the same time, it is crucial to complete a crossing programme on time to avoid differences in germination times, due to the age of the seeds and so that progenies can enter the nursery and subsequently be field planted at the same time (Breure and Verdooren, 1995). This is especially the case for the tenera x tenera crossing programmes where seeds tend to have low germination rates, and if seed has been stored for more than a year.

Thus, crossing programmes need to be carefully thought through and preparations made for palm selection, pollen collection, initiation of crossing, harvesting and seed processing (see also Kelanaputra *et al.*, 2018; Setiawati *et al.*, 2018, this series).

6.2 Plotting

Plotting helps to place all seedlings into their designated position in the field trial. The plot size will depend on the trial objectives. However, a major determinant of plot size is the inter-palm spacing required for oil palm, which is from 8.50 to 9.25 m. Plot size usually varies from 12–25 palms/plot, but in general practice progenies are arranged in a plot of 4 x 4 = 16 palms. Commonly used plot sizes and layouts are shown in Fig. 6.1.

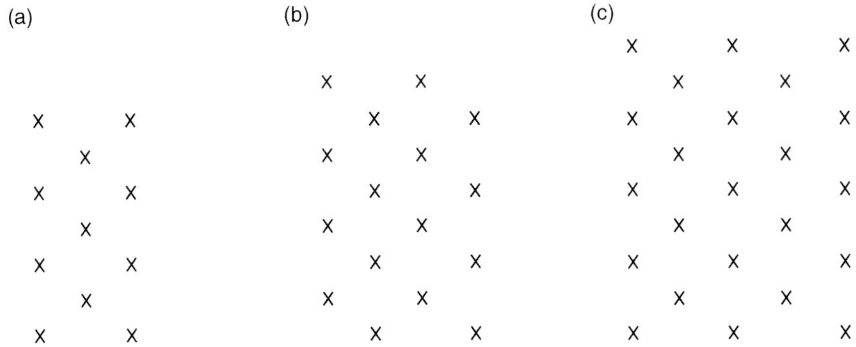

Fig. 6.1. Different plot sizes generally used in oil palm trialling where x denotes a palm position: a) 3 x 4 = 12 palms/plot; b) 4 x 4 = 16 palms/plot; and c) 5 x 5 = 25 palms/plot.

After land preparation is finished (Chapter 4) the next step is lining the area using the selected planting spacing. The making of drains is also required in the trial area at this stage (Fig. 6.2). The drains must be aligned so that they do not interfere with planting points. Ideally the drain must

Fig. 6.2. Making a drain using heavy machinery.

be in a non-planted (dead) row. Lining is also helpful in planting the cover crop, commonly *Mucuna*, which is fast-growing, tolerant to sunlight, helps to prevent erosion, fixes nitrogen and is non-competitive with mature oil palm (Corley and Tinker, 2015).

6.3 Trial Design

Oil palm breeding trials are usually planted in standard statistical designs: completely randomised or randomised complete block designs (RCBDs). Soh (1990) compared the precision of 30 different trials in Malaysia. Most trials had four or five replications of 10, 12 and 16 palms/plot, and typically could only detect differences among families of 15% or more in terms of yield. In order to detect differences of 10%, plots of 12 palms with five replications, or 16 palms with four replications, were required (on coastal soils); on the more heterogeneous inland soils, 20 palms/plot with five replications were recommended. Completely randomised and cubic lattice designs did not have any advantage over the RCBDs, nor did covariant adjustment, using the yields of neighbouring plots.

Breure and Konimor (1992) considered the optimum plot size for oil palm breeding trials. Using data from trials in Papua New Guinea and a theoretical method of inter-palm competition developed by C.J.T. Spitters, they concluded that the response to selection is greater with 8 or 16 palm plots than with the same number of palms in single palm plots. This was because inter-palm competition introduced less bias in the larger plots.

Oil palm breeding trials generally need large areas due to planting density (range from 8.5–9.5 m) and the large number of materials to be tested. Because of this, some research stations use other trial designs such as balanced incomplete block design or an alpha design. In the alpha design, each progeny is divided into small plots with only part of the treatment(s); this is thus an incomplete design, but all treatment(s) are available in each replicate which are then balanced for each replicate. For example, there can be four replicates with 16 palms/plot and if the number of progenies is in excess of 50, an alpha design trial may need 20–40 ha of land. This design does, however, reduce data bias due to variation in local conditions such as soil variation, elevation, soil fertility, etc.

6.4 Trial Layout

Trial layout can be determined once a trial design has been agreed. After land clearing, lining and drainage have been completed a base map/field layout is made, which should note information on the field condition, such as drainage, roads, etc. (Figs 6.3a, 6.3b).

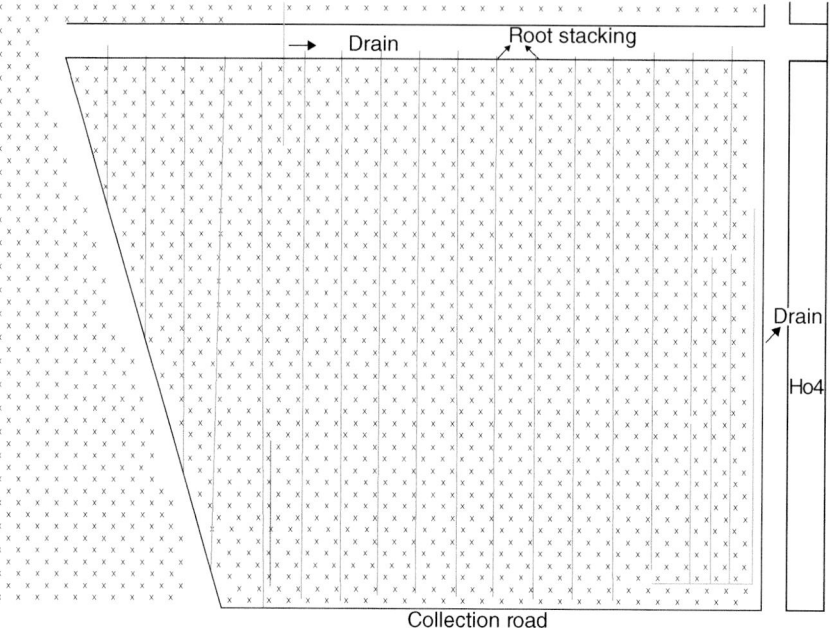

Fig. 6.3a. Basic field layout showing planting points, planting density and hand notes on other features (e.g. roads, drains, etc.). From the basic field layout, the breeder then prepares a planting layout with plots of the area (Fig. 6.3b).

6.5   Trial Planting

After the field layout is prepared, the breeder/researcher will prepare a randomised design (this is normally done in consultation with a statistician). In the example shown, a randomisation using Genstat with an alpha design has been used. After the design is set the researcher applies the design to trial layout (Fig. 6.4).

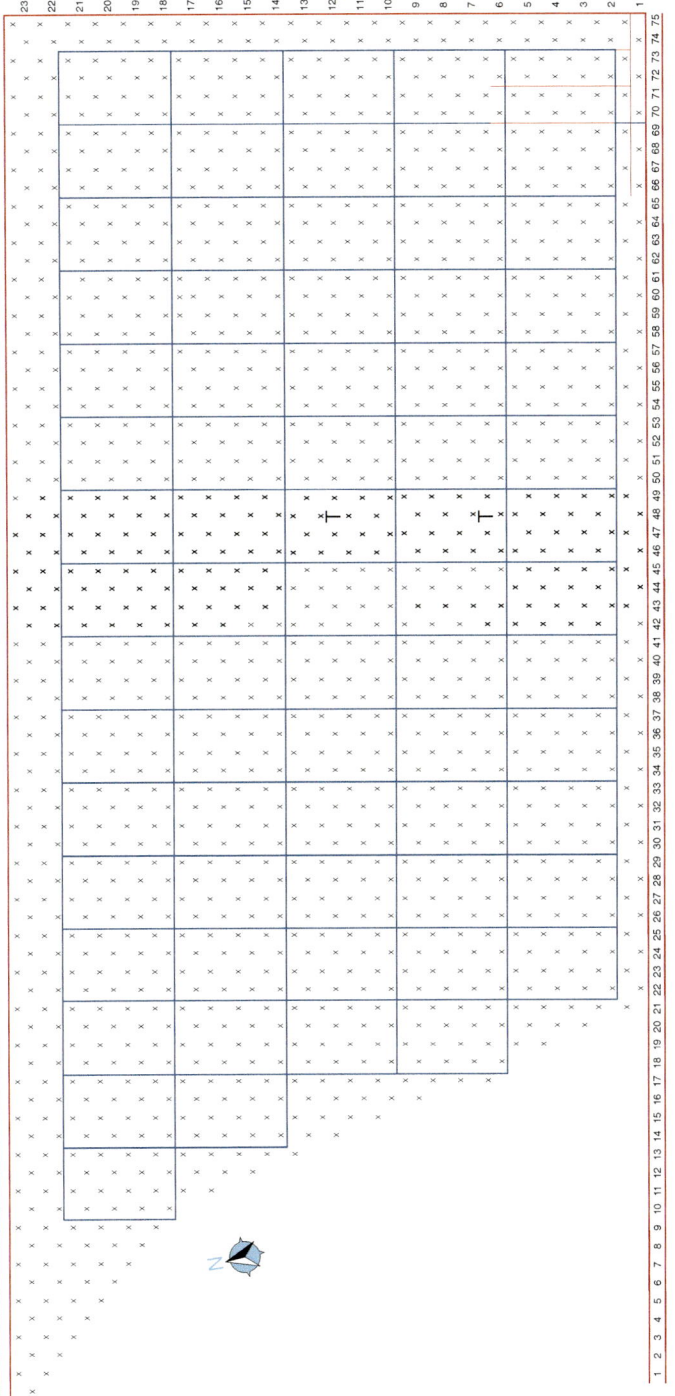

Fig. 6.3b. Plot layout showing planting positions in the planned area for a trial with a 4 x 4 plot design.

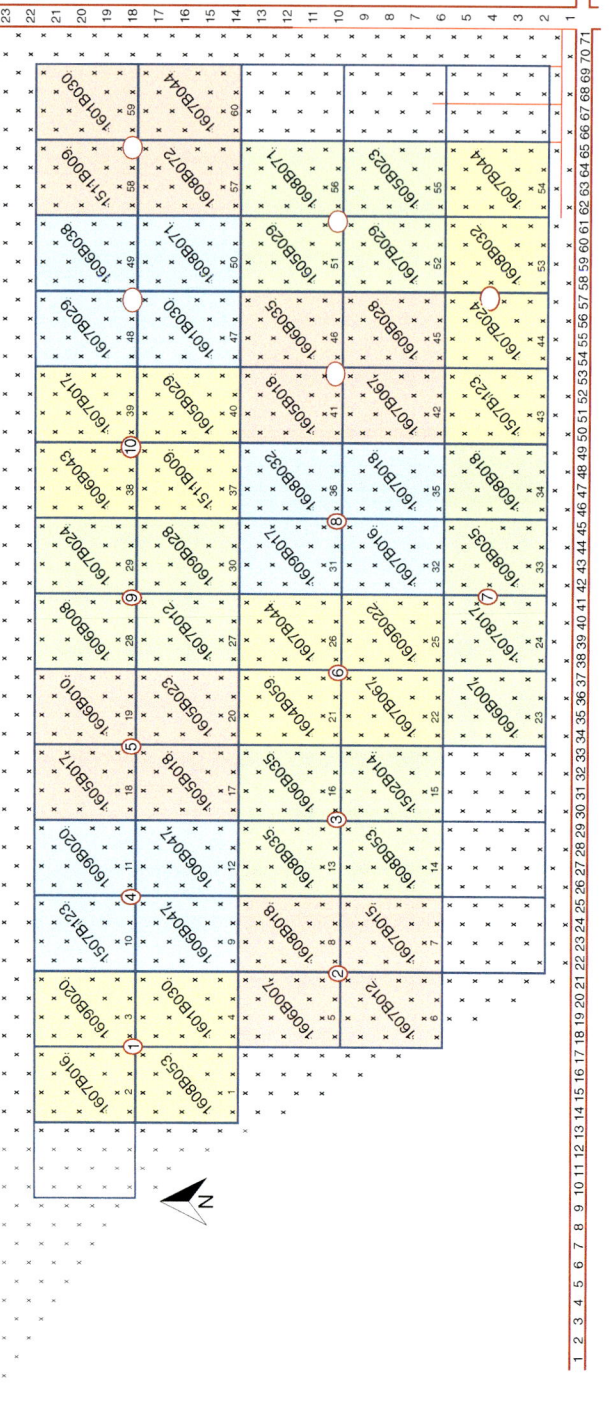

Fig. 6.4. Trial layout using an alpha (balanced incomplete block) design with plot size 4 x 4 for field planting.

References

Breure, C.J and Konimor, J. (1992) Parent selection for oil palm clonal seed garden. Rao, V., Henson, I.E. and Rajanaidu, N. (eds). *Proceedings of the 1990 ISOPB Workshop on Yield Potential in Oil Palm*, Palm Oil Research Institute, Malaysia, Kuala Lumpur, pp. 122–144.

Breure, C.J. and Verdooren, L.R. (1995) *Guidelines for Testing and Selecting Parent Palms in Oil Palm. Practical Aspects and Statistical Method*. ASD Costa Rica Publication no 9, Costa Rica.

Brown, J., Caligari, P.D.S. and Campos, H.A. (2014) *Plant Breeding*, 2nd edn. Wiley Blackwell, Chichester, UK, p. 1.

Corley, R.H.V. and Tinker, P.B. (2015) *The Oil Palm*, 5th edn. World Agriculture Series. Wiley Blackwell, Hoboken, NJ, p. 639.

Kelanaputra, E.S., Nelson, S.P.C., Setiawati, U., Sitepu, B., Nur, F. *et al.* (2018) *Seed Production in Oil Palm: A Manual. Techniques in Plantation Science*. Forster, B.P. and Caligari, P.D.S. (eds). CAB International, Wallingford, UK, p. 59.

Laksono, N.D., Setiawati, U., Nur, F., Rahmaningsih, M., Anwar, Y., Rusfiandi, H., Sembiring, E.H., Forster, B.P., Subbarao, A.S. and Zahara, H. (2019) *Nursery Practices in Oil Palm: A Manual. Techniques in Plantation Science*. Forster, B.P. and Caligari, P.D.S. (eds). CAB International, Wallingford, UK.

Rahmaningsih, M., Virdiana, I., Bahri, S., Anwar, Y., Forster, B.P. and Breton, F. (2018) *Nursery Screening for Ganoderma Response in Oil Palm Seedlings: A Manual. Techniques in Plantation Science*. Forster, B.P. and Caligari, P.D.S. (eds). CAB International, Wallingford, UK, p. 69.

Setiawati, U., Sitepu, B., Nur, F., Forster, B.P. and Dery, S. (2018) *Crossing in Oil Palm: A Manual. Techniques in Plantation Science*. Forster B.P. and Caligari, P.D.S. (eds). CAB International, Wallingford, UK, p. 50.

Soh, A.C. (1990) Oil palm breeding: breeding into the 21st century. *Plant Breeding Abstracts* 60: 1437–1444.

Wening, S., Wilkinson, M.J., Juhyana, J. *et al.* (2009) *Genetic Relationship Studies of Sumatra Bioscience Oil Palm Germplasm Using ISSR and AFLP Profiles*. IOPC 2009, Kuala Lumpur.

Recording

7

Abstract

Careful recording is an important component of field trialling. In addition to the target traits of interest, all conditions that appear during the trial should be recorded as well as anything unusual or noteworthy. All the data are used in assessing the value of the materials being tested. Recording methods may differ among oil palm research stations. However, some methods are commonly used and these are described in this chapter. Scored traits include yield and yield components (including bunch and oil analysis), growth (height increment), occurrence of common pests, diseases and physiological disorders, nutrient deficiencies and toxicities. New traits include those that can be exploited in developing mechanical harvesting systems in oil palm (fruit colour, long bunch stalks and late fruit abscission or the opposite – total fruit abscission).

7.1 Yield Recording

Yield recording includes measuring and weighing fruit bunches and any loose fruit. Bunches are weighed in the field at harvest using a spring balance attached to a tripod (Fig. 7.1). Loose (detached) fruit on the ground around the palm are also gathered, counted and weighed.

Yield recording is carried out during the normal harvesting rounds, usually every 7 to 10 days. Five years of yield recording is usually needed to provide yield stability data, which are mandatory for official progeny testing of palms for commercial seed production.

Fig. 7.1. Weighing bunches using a spring balance attached to a tripod.

7.2   Growth Measurements

Frond marking

Frond marking is carried out using paint. The youngest frond with fully expanded leaflets is marked. Marking then proceeds by determining the direction of the frond spiral, left or right. The direction of the spiral can be determined by the position of the inflorescence relative to the frond – for example, left-handed spirals have right position of the inflorescence, the opposite for right-handed spirals (Fig. 7.2).

Leaf area

Counting spines and measuring frond length
Counting the number of spines is carried out on one side of the petiole (frond stalk). Select the side of the rachis with the most spines. Counting starts at the base of the frond up to the fused leaflets at the apex (Fig. 7.4). Counting is easily done using a hand-counter.

Petiole cross-section area
This is assessed by determining the thickness and width of the petiole. Steps in measuring petiole cross-section are given in Fig. 7.5.

Length x width of leaflets
Length x width of leaflet is measured using a set of 10 leaflets (20 leaflets). These are taken from the region of 3/5 (Fig. 7.3). This region can be determined by 0.6 x frond length (see Fig. 7.6). From the sample of

Fig. 7.2. Determining the direction of frond spiral: a) left spiral; b) right spiral.

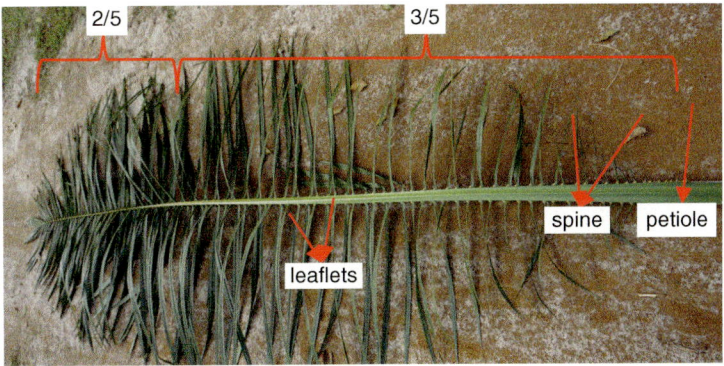

Fig. 7.3. Leaf measurements.

Fig. 7.4. a) Frond length measurement. Basal spine reference point to start measurement (right); counting spines on the side with the most spines (left).

Fig. 7.5. Measuring thickness and width of petiole with a caliper.

Fig. 7.6. Frond area taken by leaflets.

20 leaflets, the 10 longest leaflets (5 each side) are used in measuring length and width (Figs 7.7–7.9).

Trunk measurements

Height increment
Frond marking is critical in determining height increment with time. Marked fronds are used as reference points to measure the distance between the first frond at 30 months after planting with the same frond at 42 months after planting (a year later). A second measurement is also standard practice, i.e. between the first frond at 42 months after planting and the first frond at 66 months after planting (and interval of two years).

Trunk diameter
Trunk diameter is normally measured at 150 cm above the ground using callipers (Fig. 7.10).

Fig. 7.7. Selecting the longest leaflets to measure.

Fig. 7.8. Measuring leaflet length.

Fig. 7.9. Measuring leaflet width.

7.3   Palm Disorder Censuses

Crown disease

Crown disease is not a disease but a physiological disorder. It commonly occurs after transfer to the field and often the palm recovers within 35 months (Breure and Soebagjo, 1991). Crown disease surveys are carried out by walking through the plantation and assessing and scoring the condition of the palms. The value of the score is based on the number of fronds that show symptoms. There are five criteria of scoring; 1 is for the lowest number of diseased fronds and 5 means all the fronds are infected (Fig. 7.11).

Oryctes infestations

The rhinoceros beetle, *Oryctes rhinoceros* is the most important insect pest of young oil palms in Southeast Asia (Fig. 7.12). Thus, monitoring pest activity should be standard practice (e.g. by annual surveys). Assessment is based on the number of beetle-damaged fronds.

Ganoderma infections

Basal stem rot, caused by *Ganoderma boninense* is the most important disease of oil palm in Southeast Asia. The production of resistant varieties is a major oil palm breeding objective. Disease surveys are normally conducted every three months after planting.

   Oryctes rhinoceros and *Ganoderma* are the most destructive organisms for oil palm in Southeast Asia. *Fusarium* wilt is a major disease in Africa.

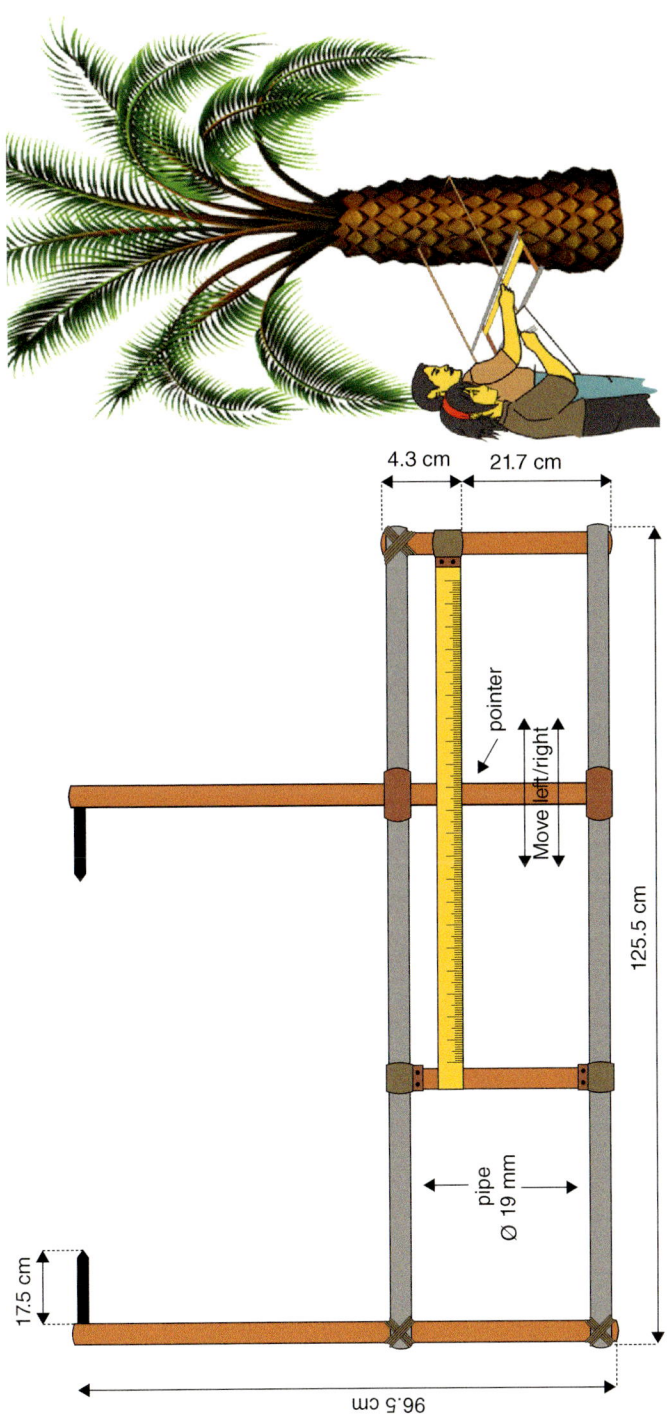

4.3 cm 21.7 cm

pointer

Move left/right

125.5 cm

pipe
Ø 19 mm

17.5 cm

96.5 cm

Fig. 7.10. Tools and methods for measuring trunk diameter.

Fig. 7.11. Crown disease: a) Score 1; b) Score 3; c) Score 5.

Fig. 7.12. Symptom of *Oryctes rhinoceros* infestations.

Other surveys

Other symptom surveys (Figs 7.13–7.15), normally carried out simultaneously on an annual basis, include:

- boron (B) deficiency
- magnesium (Mg) deficiency
- potassium (K) deficiency
- upper stem rot (USR)
- white stripe
- retarded growth
- dead palms
- sterile palms (no fruit)
- supply palm.

Fig. 7.13. Magnesium deficiency.

Fig. 7.14. Potassium deficiency.

Fig. 7.15. Boron deficiency.

7.4 Recording schedules

Timeline activities for recording are shown in Table 7.1.

Table 7.1. Times of recording.

Activity	Months After planting											
	0	12	18	24	30	36	42	48	60	66	72	84
Planting trial												
Crown disease surveys												
Annual tree surveys												
Resupplying palms												
Marking frond 1												
Yield recording												
Trunk diameter												
Leaf area												
Height increment												

7.5 Bunch and Oil Analysis

A brief description of bunch and oil analysis is given here as details of these activities are provided in Widodo *et al.* (2019, this series). Bunch analysis is an important parameter used by oil palm breeders to analyse the oil yield potential (e.g. quantity and quality comparison of bunches; between bunches within an oil palm; between oil palms; or between different progenies). Obtaining high fresh fruit bunch (FFB) production in oil palm plantations per ha area per year, and high oil extraction rates from the bunch, are the main goals of growers to maximise oil yield.

Bunch analysis mostly follows the methods of Blaak *et al.* (1963), modified by Rao *et al.* (1983) and Junaidah *et al.* (2011). The analysis involves field and laboratory operations. Activity in the field is simply bunch sampling. The bunches are then taken to the laboratory for various analyses. Activities in the bunch laboratory include:

- bunch physical analysis
- fruit sampling
- nut analysis
- oil analysis
- recording.

Blaak *et al.* (1963) divided the analysis laboratory into six sections: reception of bunches; fruit form determination; weighing the bunches; separation of spikelets from the stalk and spikelet sampling of the bunch; picking fruit from the spikelets, weighing empty spikelets and weighing the fruit; sampling fruit, scraping fruit, weighing, drying and cracking nuts, and weighing kernels; oil analysis; and recording.

Bunch sampling is done in the field (in the trial block). Either a chisel or a sickle is used according to the oil palm height or age. Harvesting is carried out according to the standard operating procedures (SOPs) of the estate, with specific interval times between each round of harvesting (7–10 days). During bunch sampling, observations are made of the status of bunch ripeness (e.g. bunch colour, number of loose fruits on the ground before and after harvesting). These will affect the recoverable oil yield and quality. For example, oil to bunch and free fatty acid content. Good practices are essential to maximise and standardise data capture for oil yield and quality.

Identification of the bunch should include details of the bunch sample (e.g. trial number or block number, palm number, row number, progeny, planting year or planting materials). The bunch sample, loose fruits and the identification label should be placed in a closed gunny sack and then transferred immediately to the bunch laboratory. During transport the bunch sample should be maintained in a good/protected condition to prevent damage (such as bruising or loss of fruit). Rao *et al.* (1983) also reported that use of gunny sacks reduces moisture loss and fruit damage in transit.

Bunch physical analyses are carried out to determine the bunch physical components and yield of oil in oil palm bunches after harvesting. Some bunch characteristics determined are:

- bunch sample weight
- fruit type (dura, tenera, pisifera)
- stalk weight
- total number of spikelets in the bunch
- number of fertile fruits in the bunch
- percentage of fertile FFBs
- number of parthenocarpic fruits in the bunch
- number of underdeveloped fruits in the bunch
- fruit set
- fruit weight
- percentage of mesocarp in the fruits
- percentage of mesocarp in the fruit check.

Bunch samples are fermented overnight by injecting them with ethepon solution by drilling two to four holes in the bunch in different positions in the stalk using an electric drill or by spraying a solution around the bunch. After overnight storage, the spikelets are separated from the stalk by chopping. Fruit type determination is achieved by taking three fruits randomly from the bunch. These are cut horizontally and the shell thickness is classified as thick (dura), thin (tenera) or absent (pisifera). Then, a random sample of spikelets is taken proportionate to the bunch weight (Table 7.2).

Spikelet samples are peeled to separate the fruits from the spikelets. The peeled fruit should have no calyx attached. The fruit is classified as either fertile, parthenocarpic or underdeveloped. Parthenocarpic and

Table 7.2. Spikelet sample size determined by bunch weight.

Bunch weight (kg)	Percentage sampled	Number of boxes needed
<7.5	100%	4
7.6 – 11.5	75%	3
11.6–16.5	50%	2
16.6–23.0	37.5%	1.5*
>23.0	25%	1

*Note: to get 1.5 boxes of samples, first take two boxes randomly from the original spikelet sample (50% sampling size). Then place the contents of the two boxes on the chopping table and drop them into four boxes. Next, take three of the four boxes (37.5% sampling size) and discard the other box.

underdeveloped fruit are removed from the fruit sample and only fertile fruits are processed for further analysis, being weighed to calculate the percentage of FFB.

Fruit samples are a random sample of fruits taken for further nut and oil analysis. Both nut and oil analysis require homogenous and representative sampling. Blaak *et al.* (1963) recommended a sample size of 500 g of fruit, but other laboratories commonly use 250 g, which is represented by 25–30 fruits.

Fertile fruits without any damage (bruised or chopped) are sampled using a randomisation box. The box divides the fruit sample into two portions (two container boxes). Thirty fruits are taken randomly from each box and placed onto two separate trays. The weight deviation between these two replicate trays should be less than 5%. More than 5% means the sample was not representative enough and the sampling should be repeated. Then, only one fruit sample in the tray will continue for further nut and oil analysis. Fruit samples are scraped to separate the wet (fresh) mesocarp from the nut. The scraping process needs to be performed soon after fruit samples are weighed to prevent the volatilisation of moisture in the sample. Once fruit scraping is complete, the scraped mesocarp should be weighed.

The data required for calculating various parameters in bunch physical analysis are recorded. They include:

- fruit type
- bunch weight
- stalk weight
- spikelet number
- fertile fruit number
- parthenocarpic fruit number
- underdeveloped fruit number
- fruit spikelet weight
- fertile fruit sample weight
- 30-fruit sample weight
- tray weight
- tray + wet (fresh) mesocarp weight
- number of nuts.

Nut analysis proceeds from the bunch physical analysis step after the fresh nuts are obtained. Some bunch characteristics determined in nut analysis are:

- percentage of shell in the fruit
- shell to kernel
- percentage of kernel in the fruit
- kernel weight
- nut weight
- kernel to nut ratio
- percentage of kernel in the bunch.

Fresh nuts should be oven dried at 60–105°C for seven to nine hours. The nuts are then cracked open to release the kernel. The kernels are counted and weighed. The data required for calculating various parameters in nut analysis are then recorded:

- fresh nut weight
- number of kernels
- kernel weight.

Oil analysis continues after the wet (fresh) mesocarp samples are obtained. Bunch characteristics determined in oil analyses include:

- percentage of dry mesocarp to fruit
- dry mesocarp to wet mesocarp
- percentage of oil in dry mesocarp
- percentage of oil in wet mesocarp
- percentage of mesocarp oil in the bunch (O/B).

Oil analysis of the mesocarp is determined by extracting the oil from dried mesocarp samples. Before the extraction step is carried out, the wet (fresh) mesocarp sample needs to be dried. Conventional drying is done in an oven at 60–105°C for 24 hours. The dried mesocarp is ground and then sieved to provide a homogenous sample for more efficient and easier oil extraction. Ground samples are placed in bags and subject to Soxhlet extraction with hexane. For safety, the solvent extraction should be carried out using an exhaust system in a vented laboratory (see also Chapter 2 with respect to hexane use). Once oil extraction is completed, oil yield can be calculated by the difference in mesocarp weight before and after extraction.

The data required for calculating various parameters in oil analysis are recorded:

- tray + dry mesocarp weight
- bag weight
- bag + dry mesocarp weight
- bag + fibre weight.

Bunch components obtained are:

- percentage of fertile FFB
- fruit set (FS)
- fruit weight (FWT)
- percentage of mesocarp in the fruit (M/F)
- percentage of mesocarp in the fruit check (M/FC)
- percentage of shell in the fruit (S/F)
- shell to kernel (S/K)
- percentage of kernel in the fruit (K/F)
- kernel to nut ratio (K/N)
- percentage of kernel in the bunch (K/B)
- percentage of dry mesocarp to fruit (DM/F)
- dry mesocarp to wet mesocarp ratio (DM/WM)
- percentage of oil in dry mesocarp (O/DM)
- percentage of oil in wet mesocarp (O/WM)
- percentage of mesocarp O/B.

O/B can be converted to an oil extraction rate (OER) by the conversion factor 0.855 x O/B as defined by Corley and Tinker (2015). Some laboratories multiply using a 0.85 correction factor to obtain an equivalent to the mill oil extraction rate, allowing for factory losses.

During the data analysis session, data checks are applied to all parameters to screen for outliers/anomalies (see Table 7.3).

The details of steps and calculations on bunch components are described in Widodo *et al.* (2019). Some bunch analysis data are extremely useful for oil palm breeders in making selections. They include:

- M/F
- O/DM
- S/K
- S/F
- O/B
- K/B.

Table 7.3. Data checks for dura and tenera samples.

Parameter	Dura		Tenera	
	More than	Less than	More than	Less than
F/B	85%	50%	85%	50%
FW	33 g	4.5 g	25 g	4.5 g
M/F	74%	43%	94%	64%
M/FC	100	93	100	93
DM/F	60%	23%	80%	35%
DM/WM	0.785	0.4	0.785	0.4
O/DM	95%	60%	96.80%	56.00%
S/F	46%	19%	24.30%	2.00%

7.6 Traits for Mechanical Harvesting

Along with breeding for high yield and keeping pace with market demands for commodity oil, crude palm oil (CPO) and its derivatives, oil palm breeders aim to breed for future demands and needs. Some of these are difficult to predict but it is very likely that with increasing labour costs, future oil palm plantations will be harvested mechanically – for which new traits need to be identified and bred into the crop. There are various approaches to this, and one is to go for full abscission of fruit with mechanical collection. Another possibility is for mechanical harvesting on the palm. For this there is a need to include a clear indicator of fruit ripeness (e.g. virescent), fruit retention (to reduce fruit loss, otherwise extra labour is needed to collect the loose fruit) and a long bunch stalk (to allow easy bunch removal from the palm). The determination of fruit ripeness in oil palm is normally performed by observing the colour change of the fruit skin (exocarp). Contemporary crops generally carry the nigrescens gene, or less commonly the virescens allele. In nigrescens, the fruit colour changes from black (unripe) to red when ripe, whereas in virescens the fruit colour changes from green to orange (Hartley, 1988, Fig. 7.16). The virescent colour change is more clearly observed from the ground and therefore of interest as a trait for mechanical harvesting.

Colour changes in fruit flesh (mesocarp) during maturation can also be observed by splitting the fruit longitudinally (Fig. 7.17), but the mesocarp colour of mature fruits is orange in both negrescens and virescens genotypes.

Fruit skin and flesh colour changes for nigrescens and virescens genotypes are outlined in Table 7.4.

The colour score of ripe oil palm bunches is generally taken 150 days after pollination and may be extended to 170 days (Transberger *et al.*, 2011). Phenotyping for fruit colour in trials is therefore easy; however one

Fig. 7.16. Fruit colour changes during fruit ripening in nigrescent (black to red) and virescent (green to orange) fruits.

Fig. 7.17. Fruit colour variants in oil palm.

has to wait until the palms reach maturity and this can take up to three years after seed germination. Genetic markers are now available for the nigrescens/virescens alleles and thus DNA analyses can be performed prior to trial planting. This can be done conveniently by testing DNA samples from nursery seedlings (Chapter 9).

The length of the bunch stalk is another characteristic of interest for mechanical harvesting. Long bunch stalks make it easier for harvesters to cut off ripe fruit bunches without the need to cut the frond midrib under the bunch, and aids harvesting using mechanical devices. Fruit bunch stalks are measured from the base of the stalk to the first spikelets of the bunch (Noh *et al.*, 2005). Bunch length can reach 35 cm in certain progenies (Fig. 7.18).

Currently the genetics of stalk length is little understood and there are as yet no candidate genes or gene markers that can be used in marker-assisted selection. Thus, the trait can only be scored at maturity, and this is most conveniently done when harvesting fruit bunches.

There are certain progenies which have harvesting periods of more than 170 days after anthesis. The harvest criteria are generally determined by the amount of loose fruit that falls on the ground. In these conditions the whole outer fruit is physiologically mature while the inner fruit is still immature

Table 7.4. Fruit skin and flesh colour changes (Bille *et al.*, 2014; Corley and Tinker, 2015).

Ripeness	Nigrescens	Virescens
Unripe	Black skin (exocarp)	Green skin (exocarp)
	White flesh (mesocarp)	Greenish flesh (mesocarp)
	White shell (endocarp)	White shell (endocarp)
	Gelatinous and cartilaginous kernel	Gelatinous and cartilaginous kernel
Ripe	Reddish skin (exocarp)	Orange skin (exocarp)
	Orange flesh (mesocarp)	Orange flesh (mesocarp)
	Black shell (endocarp)	Black shell (endocarp)
	Solid kernel	Solid kernel

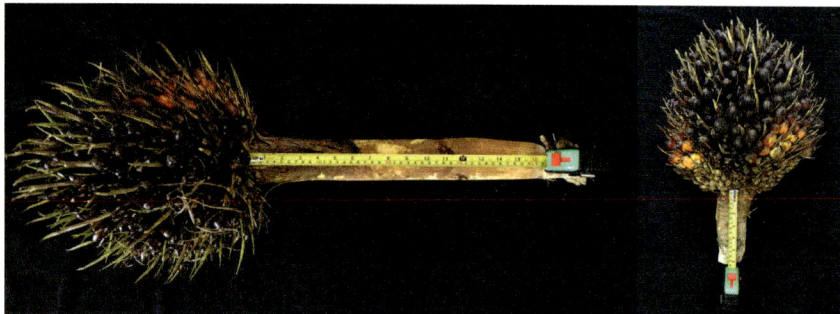

Fig. 7.18. Oil palm bunch with long stalk (left) and short stalk (right).

and the oil synthesis process can still progress (Henderson and Osborne, 1990). Thus, the potential for oil formation has not been fully achieved. There is therefore a search for germplasm with long ripening times but with delayed fruit shedding.

References

Bille, N.H., Bell, J.M., Hernild, E., Ngado, G., Nsimi-Mva, A. and Ntsefong, G.N. (2014) Morphogenesis of oil palm (*Elaeis guineensis* Jacq.) fruit in seed development. *Journal of Life Sciences* 8: 946–954, 10.17265/1934-7391/2014.12.004.

Blaak, G., Sparnaaij, L.D. and Menendez, T. (1963) Breeding and inheritance in oil palm, part II. Methods of bunch analysis. *Journal of West African Institute for Oil Palm Research*, 4: 145–155.

Breure, C.J. and Soebagjo, F.X. (1991) Factors associated with occurrence of crown disease in oil palm (*Elaeis guineensis* Jacq.) and its effect on growth and yield. *Euphytica* (54): 55–64.

Corley, R.H.V. and Tinker, P.B. (2015) *The Oil Palm*, 5th edn. Wiley-Blackwell, Chichester, UK.

Hartley, C.W.S. (1988) *The Oil Palm.*, 3rd edn. Longman, London.

Henderson, J. and Osborne, D.J. (1990) Cell separation and anatomy of abscission in the oil palm, *Elaeis guineensis* Jacq. *Journal of Experimental Botany* 41: 203–210.

Junaidah, J., Kushairi, A., Jones, B., Kho, L.E., Isa, Z.A. and Rusmin, J. (2011) Innovation for oil extraction method using NMR in bunch analysis. *International Seminar on Breeding for Sustainability in Oil Palm*, ISOPB & MPOB, Kuala Lumpur, pp. 1–18.

Noh, A., Kushairi, A., Mohd Din, A., Maizura, I., Isa, Z. A. and Rajanaidu, N. (2005) PS10: breeding populations selected for long stalk. *MPOB Information series*, ISSN 1511–7871.

Rao, V., Soh, A.C., Corley, R.H.V., Lee, C.H., Rajanaidu, N., Tan, Y.P., Chin, C.W., Lim, K.C., Tan, S.T., Lee, T.P. and Ngui, M. (1983) A critical re-examination of the method of bunch quality analysis in oil palm breeding. *PORIM Occasional Paper* 9: 7–8.

Tranbarger, T.J., Dussert, S., Joët, T., Argout, X., Summo, M., Champion, A., Cross, D., Omore, A., Nouy, B.and Morcillo, F. (2011) Regulatory mechanisms underlying oil palm fruit mesocarp maturation, ripening, and functional specialization in lipid and carotenoid metabolism. *Plant Physiology* 156: 564–584, 10.1104/pp.111.175141.

Widodo, P., Nur, F., Navisah, E., Forster, B.P. and Hasibuan, H.A. (2019) *Bunch and Oil Analysis of Oil Palm: A Manual. Techniques in Plantation Science.* Forster, B.P. and Caligari, P.D.S. (eds). CAB International, Wallingford, UK.

Ganoderma Trials **8**

Abstract

Ganoderma is the most important disease of oil palm in Southeast Asia. Resistance/tolerance to *Ganoderma* is therefore a major target for breeders and for field trialling. Methods for setting up and carrying out *Ganoderma* field trials are provided. Land preparation procedures are similar to those described in Chapter 4, with important exceptions – these are mainly to maximise disease infection in the trial and thus sanitation methods such as the removal of roots in the soil are not deployed because natural disease sources need to be maintained. Scoring disease incidence and severity in *Ganoderma* trials normally takes up to eight years from setting up the trial. As yet, little is known about the genetics of *Ganoderma* resistance and thus field trialling is the acid test in screening for improved responses to this devastating disease.

8.1   Introduction

Ganoderma boninense is a major disease that causes significant losses in the oil palm industry in Southeast Asia. The incidence of basal stem rot (BSR) caused by *Ganoderma* in oil palm plantings continues to increase, probably as a result of large-scale monoculture (Turner, 1978) and a lack of resistance. *Ganoderma* causes two types of symptoms commonly known as BSR where the base of the trunk is affected, and upper stem rot (USR), where rot occurs higher up the trunk (Durand-Gasselin *et al.*, 2005).

Recent management practices use fallow periods between crops and apply *Trichoderma* to the soil as an antagonist to *Ganoderma* (see Virdiana *et al.*, 2019, this series). An important additional component would be to plant disease-resistant oil palm. Unfortunately, there is little or no resistance in many contemporary oil palm varieties. Favoured materials such as Deli duras, originating from Malaysia and Indonesia, are more susceptible than African genotypes (Durand-Gasselin *et al.*, 2005). Other trials have revealed differences in susceptibility, indicating possible genetic resistance within populations (Idris *et al.*, 2004; Breton *et al.*, 2010). Thus, breeding for improved resistance is possible. In addition to intercrossing of elite germplasm, wild and landrace germplasm have been collected from the centre of diversity of oil palm in West Africa (Chapter 1). These new materials offer potential to contain resistance genes and crossing programmes have been initiated to introgress traits of interest into elite germplasm. The genetics of resistance (dominant/recessive, single gene/polygenes, etc.) is unknown and needs to be elucidated. Seedling testing for *Ganoderma* response is a useful early screen of germplasm and progenies (see Rahmaningsih *et al.*, 2018, this series), but currently the main test is performance in the field.

8.2   Land Preparation for *Ganoderma* Trialling

Land preparation follows the practices given in Chapter 4 with the following exceptions.

- The land selected for *Ganoderma* testing should be in an area known to have a high incidence of disease in a stand of oil palms. Aerial imaging can help select the site (Fig. 8.1).
- Since *Ganoderma* trials aim to encourage disease, there is no chipping, fallowing nor *Mucuna* planting and seedlings are planted as soon as possible after felling (Chapter 4).

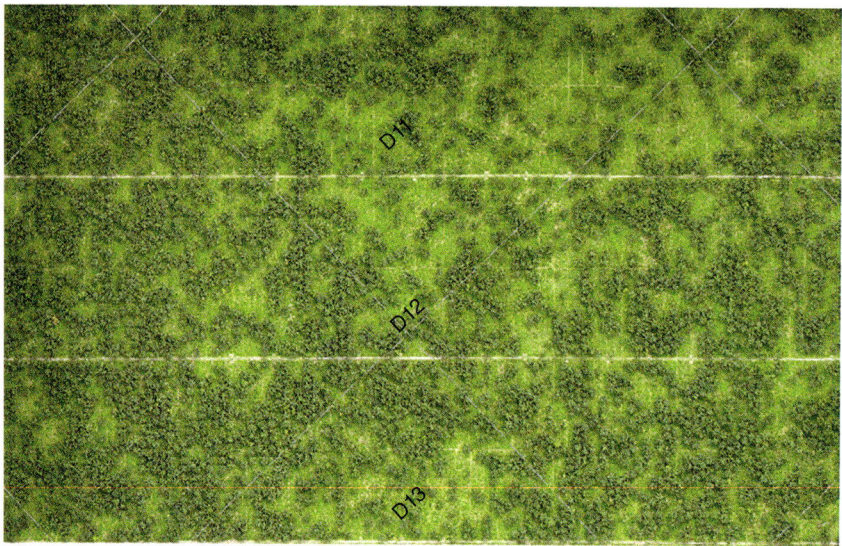

Fig. 8.1. Aerial view prior to felling of a *Ganoderma*-affected stand of oil palm, along with a proposed trial design. The bright green areas indicate loss of palms to the disease.

- There is no *Trichoderma* treatment of planting holes (see Virdiana *et al.*, 2019), unless this is used as a control plot treatment.

8.3   Set-up for *Ganoderma* Field Trial

Trial planting of a *Ganoderma* field trial is similar to a general breeding trial (Chapters 4–6). All progenies are planted randomly, using an alpha design (Chapter 6). More replication (16 replicates) with smaller-size plots (four palms per plot) is adopted by Verdant Bioscience to even out the chances of *Ganoderma* infection across the trial.

8.4   Recording of *Ganoderma* Field Trial

Besides yield recording (Chapter 7), routine *Ganoderma* surveys are conducted by workers who walk through the trial identifying diseased palms. Surveys normally start from 18 months after field planting and thereafter every 3 months until the palms are eight years old. The data recorded in a survey/census include: live palms/dead palms; the position of *Ganoderma* fruiting bodies (BSR/USR); and foliar symptom scores. The area may also be monitored by aerial photography using a drone.

Foliar scores (0 to 5) are used where 0 = no symptoms and 5 = severe symptoms (Figs 8.2–8.5). Palms are considered '*Ganoderma* infected' if the palm has died (after being recorded with *Ganoderma* symptoms) or is alive with the foliar symptoms (score 3–5) and/or disease fruiting bodies (basidiocarps) have been produced.

8.5   Data Analysis of *Ganoderma* Trial and Next Steps

The *Ganoderma* field trial is normally observed continually until the palms are felled. Therefore, the trial could last for 20 years or more. The number

Fig. 8.2. Healthy palm showing no signs of disease – score 0.

Fig. 8.3. Early symptoms of *Ganoderma* showing young leaves failing to open and the colour of leaves being pale green – score 1–2.

Fig. 8.4. Severe symptoms of *Ganoderma*. Infected palm shows dryness of fronds and rot appears – score 5.

Fig. 8.5. *Ganoderma* fruiting body (basidiocarp). The production of fruiting bodies indicates imminent palm death.

of infected palms out of the total number in a plot are normally analysed by logistic regression (binomial analysis, Purba *et al.*, 2012). The progenies are categorised or identified as resistant/tolerant when they produce low or zero percentage of infected palms in comparison with other progenies tested and standard progenies (rankings). The ranking among progenies is observed and monitored from the beginning until the end of the trial. Breton *et al.* (2010) categorised material as being resistant when infection rates of progenies were between 10–15% at the end of the planting cycle.

The selected resistant progenies will also be evaluated for yield performance and other breeding characteristics (Chapters 7 and 9). High-yielding and resistant progenies will be selected and crossed with other material to enrich the elite gene pool, or to pyramid resistance genes.

8.6 Post-*Ganoderma* Trial Activities

As mentioned above, *Ganoderma* field trials are conducted until the end of the planting cycle. Therefore, the trial is normally completed in tandem with the estate's replanting programme for the land. The area used for *Ganoderma* field trialling is usually identified as an endemic area. Consequently, the field can be re-used for subsequent *Ganoderma* field trials. Before the field is used, mapping should be conducted, for example by aerial photography, and soil sampled for DNA analysis (Chapter 1), so that *Ganoderma* populations in the soil can be assessed. Since the field is *Ganoderma*-endemic and therefore useful for further *Ganoderma* field trials, sanitation is not required. However, if the area is to be established as a plantation area, fallowing is essential, preferably followed by *Ganoderma* soil DNA testing to ensure the levels of disease are sufficiently low.

References

Breton, F., Rahmaningsih, M., Lubis, Z., Syahputra, I., Setiawati, U. *et al.* (2010) Evaluation of resistance/susceptibility level of oil palm progenies to basal stem rot disease by the use of an early screening test, relation to field observations. *Second International Seminar, Oil Palm Diseases – Advances in* Ganoderma *Research and Management*, 31 May 2010. Purba, A.R., Susanto, A., Suprianto, E., Samosir, Y. and Lubis, A.F. (eds). Indonesia Oil Palm Research Institute (IOPRI), Yogyakarta, Indonesia., Medan, Indonesia, pp. 33–62.

Durand-Gasselin, T., Asmady, H., Flori, A., Jacquemard, J.C., Hayun, Z., Breton, F. and de Franqueville, H. (2005) Possible sources of genetic resistance in oil palm (*Elaeis guineensis* Jacq) to basal stem rot caused by *Ganoderma boninense* – prospect for future breeding. *Mycophatologia* 159: 93–100.

Idris, A.S., Khushairi, A., Ismail, S. and Ariffin, D. (2004) Selection for partial resistance in oil palm to *Ganoderma* basal stem rot. *Journal of Oil Palm Research* 16: 12–18.

Purba, A.R., Setiawati, U., Rahmaningsih, M., Yenni, Y., Rahmadi, H.Y. and Nelson, S. (2012) Indonesia's experience of developing *Ganoderma* tolerant/resistant oil palm planting material. *Proceedings of 2012 International Society for Oil Palm Breeders (ISOPB) International Seminar on Breeding for Oil Palm Disease Resistance*, 21–24 November 2012. Kien, W.C. (ed.). Bogota, Colombia, Paper 6, pp. 1–22.

Rahmaningsih, M., Virdiana, I., Bahri, S., Anwar, Y., Forster, B.P. and Breton F. (2018) *Nursery Screening for* Ganoderma *Response in Oil Palm Seedlings:*

A Manual. Techniques in Plantation Science. Forster, B.P. and Caligari, P.D.S. (eds). CAB International, Wallingford, UK, p. 69.

Turner, P.D. (1978) Some aspects of natural pollination in oil palm (Elaeis guineensis). *Planter* 54, 310–328.

Virdiana, I., Rahmaningsih, M., Forster, B.P., Schmoll, M. and Flood, J. (2019) *Trichoderma: Ganoderma Disease Control in Oil Palm A Manual. Techniques in Plantation Science*. Forster, B.P. and Caligari, P.D.S. (eds). CAB International, Wallingford, UK.

Pre-trial Screening Using DNA diagnostics

9

Abstract

DNA analysis provides an opportunity to screen for certain traits prior to field planting. This is particularly important in screening for shell thickness. The shell thickness gene (*Sh*) is probably the most valuable gene in oil palm breeding and commerce. Pre-field planting screening is important as selection for dura and tenera types can be made and (sterile) pisiferas may be eliminated, thus saving on time and land space. This also allows for more robust field trialling as dura and tenera types can be separated. Nursery seedlings offer a convenient phase in which to sample leaves, extract and analyse DNA, thus selections can be made before field planting. In addition to the *Sh* gene, screening for other important traits is now possible (e.g. virescens and mantled fruit). This chapter gives an overview of pre-trial screening using currently available DNA diagnostics, but this area of work is developing quickly and set to expand in the future. Practical details of the DNA protocols involved are not given as these are in another manual in this series (Anwar *et al.*, 2019). DNA analyses can be carried out in-house or outsourced to DNA service laboratories.

9.1   Pre-trial Screening for Shell Thickness

The shell-thickness trait is economically the most important trait in oil palm. The various oil palm fruit forms (dura, pisifera and tenera) are governed by alleles of the shell thickness gene (*Sh*). The genotype of dura is *Sh/Sh* (homozygous), pisifera is *sh/sh* (homozygous) and tenera is *Sh/sh* (heterozygous). Tenera types form the commercial crop and are produced from crossing dura seed palms with pollen from pisifera males (Chapter 1). The identification of desired fruit genotypes is now a major asset to oil palm breeders (and oil palm seed producers) in determining and separating the various types, confirming legitimacy and detecting contaminants. For breeding, this saves time and space as only specific types are selected and advanced to field trialling and trial efficiency is thus increased (Babu *et al.*, 2017).

The *Sh* gene has been described by Singh *et al.* (2013). Sequence data of *Sh* and homoeologues have been exploited in developing DNA diagnostics for dura, pisifera and tenera genotypes, and this can be done from nursery seedling samples (Ooi *et al.*, 2016). *Sh* sequencing results show single nucleotide polymorphisms (SNPs), which are associated with variation for fruit type. These differences can be detected using various DNA methods, especially those involving the polymerase chain reaction (PCR) to amplify target DNA regions (e.g. high resolution DNA melt – HRM – curves as shown in Fig. 9.1). Cleaved amplified polymorphism (CAP) analysis may also be used to verify the HRM results.

The workflow for pre-trial shell thickness determination is as follows.

1. Leaf sampling of seedlings (or ramets) in the nursery.
2. DNA extraction in the laboratory.
3. Amplification of DNA with HRM PCR.
4. Shell thickness determination with HRM.
5. Verification of the HRM results with CAPs and sequencing.
6. Selection of palms in the nursery according to *Sh* genotyping.
7. Planting to the field according to trial design (see previous chapters).

DNA primers are designed that target the SNP mutation site and DNA is amplified using PCR (Qiagen, 2009). The amplified DNA is then subject to analysis by HRM. Three different melt curves, for pisifera, tenera and dura, can be detected (Fig. 9.1). HRM requires real-time PCR machines, such as the Rotor-Gene Q (Qiagen) and can be used for rapid screening of large sample sizes in high-throughput screening. One machine can handle

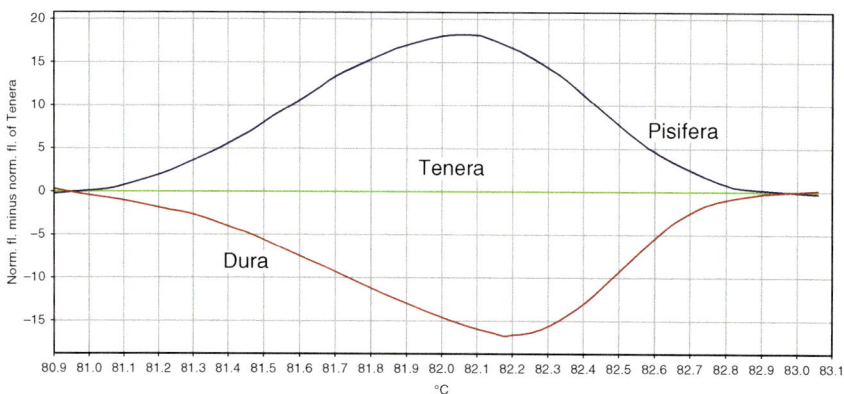

Fig. 9.1. Differential HRM curves for dura, pisifera and tenera (transformed using tenera as a baseline).

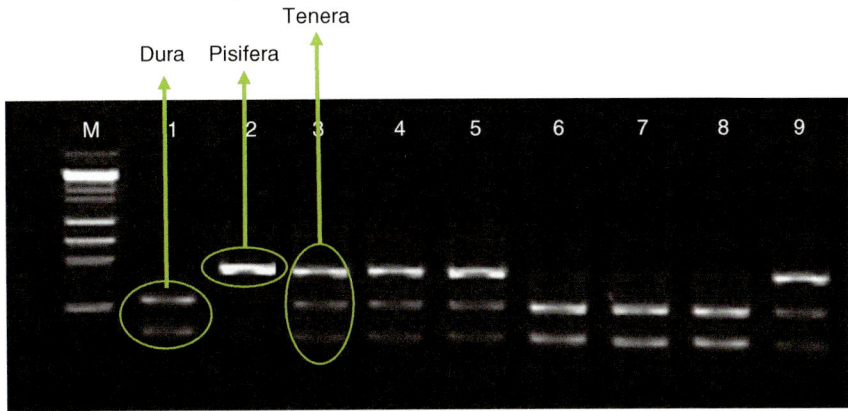

Fig 9.2. Verification of *Sh* determination with CAP fingerprints; 1–3 are positive controls of dura (two bands), pisifera (one band) and tenera (three bands); 4–9 are samples (tenera, tenera, dura, dura, dura and tenera).

36, 72 and 100 samples per run based on rotor type and is thus suited to screening individuals of large (breeding) populations.

In certain circumstances it may be necessary to verify/check the HRM data (e.g. non-specific products or post-PCR artefacts such as primer-dimers). This can be done by producing cleaved amplified products (CAPs), where DNA is cleaved (or not) depending on sequence/cleavage enzyme combinations, thus producing one (uncleaved), two or three products (Fig. 9.2).

There are many methods available for genotyping. In addition to the in-house protocols for *Sh* outlined above, *Sh* determinations can be done using outside services such as Orion Biosains (www.orionbiosains.com) where methods are based on The SureSawit *SHELL* test analysis.

9.2   Genotyping for Traits

Virescens

Oil palm fruit colour is a target for helping to enable efficient mechanical harvesting (Chapter 7). The two most common colour types are nigrescens (nig) and virescens (vir). Virescens is controlled by a single locus with vir being a dominant allele. The vir fruits are green in colour when unripe and change to bright orange when ripe, due to an absence of carotenoids in the skin (exocarp). This distinct colour change allows easy identification of ripe bunches and hence maximises yield (Jack *et al.*, 1998).

DNA diagnostic assays for vir provide opportunities to develop new and improved virescens breeding lines and commercial varieties. Like *Sh*, DNA diagnostics for vir can be done on nursery plants and selections made before transfer to field trials. Sequence data of vir is available and allows marker-assisted selection for both nigrecens and virescens alleles (e.g. by HRM analysis. Vir/nig determination can also be done using the TeraSawit™ VIR test that has been developed by Orion Genomics (http://oriongenomics.com/en/), which is a high-throughput DNA assay that not only enables differentiation of virescens from nigrescens, but also allows homozygous and heterozygous forms of virescens to be identified.

Mantled fruit

Cloning via tissue culture is commonly used in oil palm but may result in somaclonal variation such as mantling of fruit which produces abnormal flowering and fruit with little to no oil yield. This was a devastating phenomenon in early work on oil palm commercial cloning in the mid-1980s (Durand-Gasselin *et al.*, 1993), and oil palm growers remain understandably nervous about clonal material produced from tissue culture. The mantled phenotype is epigenetic in nature. Shearman *et al.* (2013) performed RNA-Seq on developing flower and fruit samples of normal and mantled oil palms to characterise their transcriptomes. They found that many genes are differentially expressed, including disruption in several pathways that may cause the mantled phenotype, such as genes involved in primary hormone responses, DNA replication and repair, chromatin remodelling and a gene involved in RNA-mediated DNA methylation. Thus, the genetic determination causing mantled flowers was identified. Characterisation of the mantled gene revealed the molecular mechanism associated with the mantled phenotype in oil palm clones (Ong-Abdullah *et al.*, 2015). A Karma transposon was found to be inserted into the mantled gene, resulting in DNA methylation which produces the abnormal mantled phenotype. This DNA methylation can now be detected and thus nursery materials can be screened, and removed, prior to field planting (Lakey *et al.*, 2016).

References

Anwar, Y., Apriyanti, D., Ciomas, J., Forster, B.P. *et al.* (2019). *DNA Analysis in Oil Palm Breeding and Seed Production: A Manual. Techniques in Plantation Science*. Forster, B.P. and Caligari, P.D.S. (eds). CAB International, Wallingford, UK, in preparation.

Babu, B.K., Mathur, R.K., Kumar, P.N., Ramajayam, D., Ravichandran, G., Venu M.V.B. *et al.* (2017) Development, identification and validation of CAPS marker for SHELL trait which governs dura, pisifera and tenera fruit forms in oil palm (*Elaeis guineensis* Jacq.). *PLoS ONE* 12(2): e0171933. doi:10.1371/journal.pone.0171933.

Durand-Gasselin, T., Duval, Y., Baudouin, L., Maheran, A.B., Konan, K. and Noiret, J.M. (1993) Description and degree of the mantled flowering abnormality in oil palm (*Elaeis guineensis* Jacq) clones produced using the orstom-CIRAD procedure. *Proceedings of the 1993 ISOPB International Symposium on Recent Developments in Oil Palm Tissue Culture and Biotechnology.* Rao, V., Henson, I.E. and Rajanaidu, N. (eds). Kuala Lumpur, Malaysia, 24–25 Sept., 00. 48–63.

Jack, P.L., James, C., Price, Z., Rance, K., Groves, L., Corley, R.H.V., Nelson, S.P.C. and Rao, V. (1998) Application of DNA markers in oil palm breeding. *Proceedings of the 1998 International Oil Palm Congress Commodity of the Past, Today and Future.* Jatmika, A. (ed.). Indonesian Oil Palm Research Institute, Medan, Indonesia. pp. 315–324

Lakey, N., Singh, R., Ong-Abdullah, M., Low, E.T.L., Ooi, L.C., Nookiah, R. *et al.* (2016) Translating the oil palm genome information into precision agriculture breeding practice. *Proceedings of the International Seminar on Oil Palm Breeding and Seed Production and Field Visits*, p. 36. Malaysian Palm Oil Board, Malaysia.

Ong-Abdullah, M., Ordway, J.M., Jiang, N., Ooi, S.E., Kok, S.Y., Sarpan, N. *et al.* (2015) Loss of Karma transposon methylation underlies the mantled somaclonal variant of oil palm. *Nature* 525: 533–537.

Ooi, L.C., Low, E.T.L., Ong-Abdullah, M., Nookiah, R., Ting, N.C., Nagappan, J. *et al.* (2016) Non-tenera contamination and the economic impact of SHELL genetic testing in the Malaysian independent oil palm industry. *Frontiers in Plant Science* 7:771.

Qiagen (2009) *Type-it HRM PCR Handbook*. Qiagen, Germany.

Shearman, R.J., Jantasuriyarat, C., Sangsrakru, D., Yoocha, T., Vannavichit, A. *et al.* (2013) Transcriptome analysis of normal and mantled developing oil palm flower and fruit. *Genomics* 101: 306–312.

Singh, R., Low, E.T.L., Ooi, L.C.L., Ong-Abdullah, M., Ting, N.C., Nagappan, J. *et al.* (2013) The oil palm SHELL gene controls oil yield and encodes a homologue of SEEDSTICK. *Nature* 500: 340–344.

Index

Page numbers in **bold** type refer to figures and tables.

CABI – who we are and what we do

This book is published by **CABI**, an international not-for-profit organisation that improves people's lives worldwide by providing information and applying scientific expertise to solve problems in agriculture and the environment.

CABI is also a global publisher producing key scientific publications, including world renowned databases, as well as compendia, books, ebooks and full text electronic resources. We publish content in a wide range of subject areas including: agriculture and crop science / animal and veterinary sciences / ecology and conservation / environmental science / horticulture and plant sciences / human health, food science and nutrition / international development / leisure and tourism.

The profits from CABI's publishing activities enable us to work with farming communities around the world, supporting them as they battle with poor soil, invasive species and pests and diseases, to improve their livelihoods and help provide food for an ever growing population.

CABI is an international intergovernmental organisation, and we gratefully acknowledge the core financial support from our member countries (and lead agencies) including:

Ministry of Agriculture People's Republic of China

Australian Government
Australian Centre for International Agricultural Research

Agriculture and Agri-Food Canada

Ministry of Foreign Affairs of the Netherlands

Schweizerische Eidgenossenschaft
Confédération suisse
Confederazione Svizzera
Confederaziun svizra
Swiss Agency for Development and Cooperation SDC

Discover more

To read more about CABI's work, please visit: **www.cabi.org**

Browse our books at: **www.cabi.org/bookshop**, or explore our online products at: **www.cabi.org/publishing-products**

Interested in writing for CABI? Find our author guidelines here: **www.cabi.org/publishing-products/information-for-authors/**